Florian Rieger

Work-hardening of dual-phase steel

Schriftenreihe
Kontinuumsmechanik im Maschinenbau
Band 7

Karlsruher Institut für Technologie (KIT)
Institut für Technische Mechanik
Bereich Kontinuumsmechanik

Hrsg. Prof. Dr.-Ing. habil. Thomas Böhlke

Eine Übersicht aller bisher in dieser Schriftenreihe erschienenen Bände finden Sie am Ende des Buchs.

Work-hardening of dual-phase steel

by
Florian Rieger

KIT Scientific Publishing

Dissertation, Karlsruher Institut für Technologie (KIT)
Fakultät für Maschinenbau
Tag der mündlichen Prüfung: 15. Mai 2015

Impressum

KIT Scientific Publishing

Karlsruher Institut für Technologie (KIT)
KIT Scientific Publishing
Straße am Forum 2
D-76131 Karlsruhe

KIT Scientific Publishing is a registered trademark of Karlsruhe Institute of Technology. Reprint using the book cover is not allowed.

www.ksp.kit.edu

Print on Demand 2016

ISSN 2192-693X
ISBN 978-3-7315-0513-6
DOI 10.5445/KSP/1000054047

Work-hardening of dual-phase steel

Zur Erlangung des akademischen Grades

Doktor der Ingenieurwissenschaften

der Fakultät für Maschinenbau,
Karlsruher Institut für Technologie (KIT)

genehmigte
Dissertation
von

Dipl.-Ing. Florian Rieger

Tag der mündlichen Prüfung 15.05.2015
Prof. Dr.-Ing. habil. Thomas Böhlke (Hauptreferent)
Prof. Dr. rer. nat. habil. Dr. h.c. Juri Estrin (Korreferent)
Prof. Dr.-Ing. Martin Heilmaier (Vorsitz)

Zusammenfassung

Aufgrund ihrer guten Kaltumformungseigenschaften werden niedrig legierte Dualphasen-Stähle (DP-Stähle) häufig im Automobilbereich eingesetzt. Die kompositartige Mikrostruktur der Stähle besteht aus einer duktilen, ferritischen Matrix und $10 - 60\,\text{vol.\%}$ martensitischen Einschlüssen. Die folgenden drei Themen wurden untersucht: Einfluss der Korngrößenverteilung, Plastizität auf Kornebene und Herleitung eines neuartigen Materialmodells.

Im Kontext des korngrößenabhängigen Fließverhaltens von Mehrkristallen wird eine logarithmische Normalverteilung zusammen mit korngrößenabhängigem plastischem Verhalten angenommen. Vor allem für kleine Korndurchmesser um ein Mikrometer, führen Korngrößenverteilungen zu einer deutlichen Reduktion der Festigkeit. Die numerischen Ergebnisse des Homogenisierungsverfahrens können durch einen einfachen analytischen Ausdruck gut genähert werden.

Das mikromechanische Verhalten von DP-Stählen wird zusätzlich durch Vollfeld-Simulationen einer kristallplastischen ferritischen Matrix und martensitischen Einschlüssen vom von Mises Typ untersucht. Eine höhere Korngrenzenabdeckung mit Martensit führt zu einer höheren durchschnittlichen Anfangsversetzungsdichte nach dem Abschrecken. Für Zugdeformationen über $\sim 10\,\%$ stellt sich korngrößenabhängig ein umgekehrter Effekt ein: eine geringere oder langsamere Verfestigung für höhere Martensitabdeckungen.

Das erarbeitete Materialmodell für DP-Stähle basiert auf einem Homogenisierungsverfahren vom Hashin-Shtrikman Typ. Das neuartige Modell realisiert die gemittelte wechselseitige Beeinflussung von Ferrit und Martensit durch weitreichende Spannungen, welche durch plastische Inkompatibilität verursacht werden. Das vorgeschlagene Materialmodell vereint die Arbeiten von Ashby (1970) und Brown and Stobbs (1971a), um das Ferritverhalten zu simulieren. Basierend auf dem Model für Verbundstrukturen von Mughrabi (1983), wird die Wechselwirkung der Gefügebestandteile durch kinematische Verfestigung abgebildet. Das Model reproduziert das gemittelte makroskopische und mikroskopische Verhalten des untersuchten DP600-Stahls, sowie die Entwicklung der weitreichenden Spannungen. Darüber hinaus stimmt das entwickelte Modell gut mit experimentellen Daten für eine Reihe von DP-Stählen aus der Literatur überein.

Summary

Exhibiting good mechanical properties for cold-sheet forming, low-alloyed dual-phase (DP) steels are nowadays widely used for automotive applications. The composite-like microstructure of DP steels is composed of a low-carbon ductile ferrite-matrix and $10-60\,\mathrm{vol.\%}$ hard martensitic inclusions. A nonlinear mean-field model and full-field finite-element simulations are applied to investigate three major topics: the influence of grain-size distribution, grain-level plasticity and derivation of an original material-model.

The plastic behavior of polycrystals is assumed to be grain-size dependent in this work. The distribution of grain-sizes is taken to be lognormal. It is found that grain-size dispersion leads to a decrease of the material strength, in particular for small mean diameters around one micron. The numerical results from the mean-field model are confirmed notably well by means of a simple analytical expression.

The micromechanical behavior of DP steels is investigated by full-field RVE simulations with a crystal-plasticity based ferrite-matrix and von Mises-type martensite inclusions. To examine the martensite influence, full-field simulation results of DP steels have been compared to an RVE in which martensite is substituted by ferrite. After quenching, a higher grain-boundary area covered by martensite facilitates an increased average dislocation-density. For uniaxial deformations above $\sim$10%, however, the grain-size dependent relation reverses. With more surrounding martensite, the local crystal-plasticity material-model exhibits hardening at a slower rate.

A nonlinear mean-field model of Hashin-Shtrikman type is employed as framework for the original material-model for DP steels. The model incorporates the interaction of ferrite and martensite via incompatibility-induced long-range stresses in an averaged sense. The proposed model combines works of Ashby (1970) and Brown and Stobbs (1971a) to simulate the ferrite behavior. Based on the composite model of Mughrabi (1983), the constituent interaction is incorporated by a kinematic-hardening contribution. The model reproduces the averaged macroscopic, microscopic, and long-range stress behavior for the investigated DP600 steel. Furthermore, the model features good agreement with experimental data for a range of different DP steels from the literature.

Danksagung

Die vorliegende Arbeit entstand während meiner Zeit als Doktorand am Institut für Technische Mechanik – Lehrstuhl für Kontinuumsmechanik. Das von mir bearbeitete Promotionsprojekt A12 des Graduiertenkollegs 1483 wurde von der Deutschen Forschungsgemeinschaft finanziert.

Für die große Unterstützung während der Promotion möchte ich meinen tief empfundenen Dank gegenüber allen beitragenden Personen zum Ausdruck bringen.

Mein besonderer Dank gilt meinem Doktorvater Professor Thomas Böhlke. Er stand mir während dieser herausfordernden Promotion mit seinem Fachwissen zur Seite und fand dabei auch Zeit für persönliche Anliegen. Mein Dank gilt auch meinem Zweitgutachter Professor Juri Estrin, der mir den fachlich und persönlich wertvollen Forschungsaufenthalt an der Monash University ermöglicht hat. Ohne die Förderung des Auslandsaufenthaltes durch das Karlsruhe House of Young Scientists (KHYS) wäre dieser nicht möglich gewesen. Darüber hinaus bedanke ich mich bei Professor Martin Heilmaier für den Vorsitz der Promotionsverteidigung.

Moritz, Simone und Rene möchte ich für wertvolle Messdaten und viele Diskussionen während unserer Zusammenarbeit danken. Auch viele andere Kollegen – Rudolf, Vedran, Jan, Daniel, Eric und Susanne – haben fachlich und persönlich zu dieser Arbeit beigetragen. Besonders Ute und Helga waren durch ihre Unterstützung in “allen anderen” Dingen unverzichtbar.

Am meisten jedoch bedanke ich mich bei meiner Familie und meinen Eltern die mich während der Promotion jeweils mit jeder möglichen Unterstützung bedacht haben.

Für Reta, Vincent und Paulina

Contents

Chapter 1

Introduction

Dual-phase steels (DP) are nowadays widely used for automotive applications due to their weight reduction potential and good formability. Crash-relevant structural vehicle-parts like the A-pillar are, for example, deep-drawn from dual-phase steel sheets. The composite-like microstructure of DP steels consists of a low-carbon, ductile ferrite matrix and of hard martensitic inclusions $10 - 60\,\mathrm{vol.\%}$.

DP steels exhibit a low initial yield-strength with subsequent continuous yielding. After a high amount of work-hardening, the high initial work-hardening rate drops of at high tensile strengths. Uniform and total elongation are comparably high regarding other steels with similar yield-strengths (Erdogan and Tekeli, 2002). This facilitates good mechanical properties of DP steels for cold-sheet-forming. The mechanical properties can be adjusted within a wide range despite the low amount of alloy contributions. This makes DP steels a very cost effective material.

With an increasing martensite volume fraction c_{M}, DP steels generally show an incline in tensile strength, while the formability (uniform and total elongation) is reduced. Additionally, the microstructure of DP steel strongly influences its behavior (Balliger and Gladman, 1981). For example, finer dispersed martensite-islands lead to an increased work-hardening rate, and banded martensite causes anisotropic properties.

1.1 Initial state of dual-phase steels after production

The normal production process for DP steels may consist of initial hot-rolling or cold-rolling steps to decrease the grain size. Afterward, the material is intercritically annealed above the Ac1 temperature to create an austenite-ferrite microstructure. The subsequent quenching step triggers the austenite-martensite phase transformation and is often followed by hot galvanization or general tempering (Kuziak et al., 2008). The body-centered tetragonal martensite lattice occupies more specific volume than face-centered cubic austenite. Effectively, this leads to a volume expansion where surrounding ferrite grains deform plastically to maintain compatibility. This mechanism creates an inhomogeneous and pre-deformed material state at the end of the production (Calcagnotto et al., 2010a). Crystallographic orientations can be measured by electronic backscatter diffraction (EBSD) in ferrite. Martensite, however, is heavily distorted due to the volume-expansion being restricted by the ferrite matrix. This leads to a low attainable image quality for common EBSD measurements, often rendering this method unfeasible for dual-phase steels.

The overall carbon concentration of DP steels $(0.05 - 0.2\,\mathrm{wt.\%})$ enriches in austenite during the annealing process. Due to the rapid phase-transformation the carbon content of martensitic inclusions, generally, remains high within values of $\sim 0.2 - 0.7\,\%$. Carbon content is the main influence on the martensite yield-strength (Krauss, 1999; Delince et al., 2007). Similar to interstitial-free steels, ferrite exhibits carbon concentrations around the solubility level of carbon $(0.001 - 0.03\,\mathrm{wt.\%})$ (Resende and Bouvier, 2013). Therefore, effects of the ferrite matrix in DP steels are often explained by mechanisms that were observed in interstitial-free steel (e.g., Dillien et al., 2010).

The inhomogeneous initial material state and its effects on the work-hardening have been experimentally measured and investigated by simulations. Using micro-indentation tests, Kadkhodapour et al. (2011) and Tsipouridis et al. (2011) identified different gradients in the material hardness near grain and phase boundaries. Compared to ferrite-ferrite grain boundaries (FF), the ferrite-martensite phase boundaries (FM) have been shown to posses a significantly higher impact on dislocation motion. The works from Calcagnotto et al. (2010a) and Ramazani et al. (2013) estimate the density of geometrically-necessary dislocations (GND) and find similar quantitative results for initial dislocation-density gradients. Both works highlight the different extent to which ferrite-martensite and ferrite-ferrite boundaries influence the surrounding material behavior. In the cited works, as well as in Liedl et al. (2002), the initial GND gradient was found to increase the initial material-strength due to the increased dislocation-density in ferrite. Dillien et al. (2010) found a clearly visible initial gradient in the lattice orientation near ferrite-martensite boundaries and estimated the maximum excess dislocation-density after quenching to be $6 \cdot 10^{13}\,\mathrm{m}^{-2}$ near ferrite-martensite interfaces.

1.2 Work-hardening behavior of dual-phase steels

Many researchers find linear relationships for mechanical properties of DP steels with different volume-fractions but the same principal processing routes (Davies, 1978; Chen and Cheng, 1989; Mazinani and Poole, 2007a; Chakraborti and Mitra, 2007). However, extrapolations of these linear relationships to pure ferrite and to higher martensite contents result in erroneous estimations, indicating the limitations of these approaches (Chakraborti and Mitra, 2007). Additionally, the linear slope often varies for different production routes and absolute values of, e.g., the yield strength, depend on the global carbon content as well (Chen and Cheng, 1989). A possible explanation might lie in the observation that similar production routes (if, e.g., only durations or temperatures are altered) result in similar microstructures. In other words, if only durations or temperatures are altered, only the martensite volume-fraction but not the distribution or shape changes. Fundamentally different production routes, where, e.g., rolling takes place before or during annealing, produce fundamentally different microstructures. The complex influences of morphology and the inhomogeneous initial material-state on the work-hardening behavior of DP steels have been widely investigated and several phenomena can be identified in the literature:

The dislocation evolution in DP steels during external loading has been accounted for, e.g., in the work of Korzekwa et al. (1984). They find an initial dislocation-density gradient near martensite islands, while the interior of ferrite grains exhibits low dislocation-densities. This gradient spreads into the interior of ferrite grains. Deformation is generally higher near martensite islands during the complete loading process. For high strains of about $50\,\%$, ferrite-ferrite boundaries could not be identified anymore, and the ferrite dislocation-density became relatively homogeneous at values of $4 \cdot 10^{14}\,\mathrm{m}^{-2}$. At higher strains, the dislocation patterns (dislocation cell sizes) become similar near ferrite-martensite and ferrite-ferrite boundaries. Dillien et al. (2010) show that the orientation-incompatibility effect of ferrite-ferrite boundaries remains observable at even lower strains. Additionally, the dislocation density near ferrite-martensite boundaries increases from $6 \cdot 10^{13}\,\mathrm{m}^{-2}$ to $1.5 \cdot 10^{14}\,\mathrm{m}^{-2}$ and the width of the influence-zone increases as well. Even for low macroscopic strains, Kapp et al. (2011) find an increased strain-heterogeneity and interrupted high-strain bands in ferrite. Generally, a high usage of the ferrite work-hardening capacity is achieved by a decreased ferrite grain-size in conjunction with finely-dispersed martensite. Both geometrical features facilitate fine shear-bands that are spread over the entire ferrite matrix (Tasan et al., 2014).

The yield-strengths martensite is higher than that of ferrite by a factor of $2-4$. Thus, stress and strain partitioning is a pronounced phenomenon in DP steels. The ductile ferrite-matrix carries the majority of deformation while the harder martensite inclusions exhibit much higher average stresses. A significant strain partitioning builds up during the initial deformation-phase. The strain partitioning between ferrite and martensite becomes constant if the martensite load is sufficiently high for plastic deformation (Byun and Kim (1993), Mazinani and Poole (2007a), Thomser et al. (2009)). Han et al. (2013) find that the microstructure morphology is the major influence on strain localization phenomena, when compared to the effect of ferrite-ferrite orientation incompatibilities.

Long-range stresses DP steels generally exhibit a distinct Bauschinger-effect, i.e., after pre-deformation, the yield strength for subsequently reversed loading is significantly lower than at the end of the forward loading. Thus, two hardening components can be identified in DP (Dillien et al., 2010): isotropic forest hardening that is observable with EBSD measurements; and hardening from dislocation patterns that induce long-range stresses. The distinction between more stable forest dislocation-patterns and unstable, ordered dislocation-patterns becomes a necessity for non-monotonic load paths (Delince et al., 2007; Rauch et al., 2011). This is particularly important for DP steels. Their complex microstructure pronounces the build-up of unstable dislocation pile-ups at ferrite-martensite interfaces (Gardey et al., 2005).

The distinct Bauschinger-effect is caused by long-range stresses from dislocation arrangements with a dipole structure (Brown and Stobbs, 1971a; Mughrabi, 2006). These long-range stresses lower the effectively applied load in ferrite and in turn increase the loading of martensite. Thus, long-range stresses induce a load-transfer from ferrite to martensite. Seyedrezai (2014) experimentally investigates the long-range stress contribution to work-hardening with respect to the microstructure. It is shown that the ferrite grain-size and the martensite distribution influence the strain incompatibility. Furthermore, the experimental results of Seyedrezai (2014) show that the martensite morphology influences the efficiency of the load-transfer form ferrite to martensite, significantly.

To the best knowledge of the author, there exists no dislocation-density based mean-field material model, that is i) simple *and* physically-sound for different DP steels and ii) incorporates important microstructural features besides the grain-size. There are several works utilizing two- or three-dimensional RVEs in conjunction with crystal-plasticity models (e.g., Chen et al., 2014). While this approach has been shown to yield good results, the computational efforts are comparably high. The inclusion of non-local effects, for example long-range stresses or gradient-effects, increase the computational costs of full-field simulations even further.

1.3 Scale separation and transition

Real materials often possess a heterogeneous microstructure. In the case of DP steels, the heterogeneity is clearly pronounced since ferrite and martensite build a composite-like structure on the micro scale, i.e., on the scale of micrometers or single grains. On the length scale of the structural part, millimeters to meters, DP steels are supposed to behave homogeneously. Scale separation is ensured by several orders of magnitude between local dislocation-interactions and structural-part dimensions. However, microscopic composition and dislocation-obstacle interactions decisively influence the macroscopic behavior. No additional mesoscopic scale is introduced in between the microscopic and macroscopic scale. Thus the mean-field approach presented in this work, is essentially a two-scale approach.

With respect to the application of micromechanics within a finite-element procedure, two different transitions between the two scales are considered. Localization, i.e., transition from macro to micro scale is accomplished by a Hashin-Shtrikman based mean-field method. Based on the macroscopic-strain input of the finite-element procedure at each integration point, the average strains of the individual phases are iteratively calculated. The homogenization, i.e., the transition from micro to macro scale, can be accomplished by using the ensemble average for the stress tensor,

$$\bar{\boldsymbol{\sigma}} := \langle \boldsymbol{\sigma} \rangle = \frac{1}{V} \int_{\mathrm{RVE}} \boldsymbol{\sigma}(\boldsymbol{x}) \, \mathrm{d}V = \sum_{\alpha=1}^{N} c_\alpha \langle \boldsymbol{\sigma} \rangle_\alpha . \tag{1.1}$$

Since the calculated ferrite and martensite behavior is assumed to be the ensemble average of all ferrite or martensite grains, the ensemble average for the macroscopic stress simplifies to two contributions. The detailed homogenization of algorithmic effective-tangent tensors for the global finite-element newton-scheme is discussed in Section 6.8.

Equation (1.1) necessitates the definition of a representative volume-element (RVE). A volume element for DP steels must exhibit several characteristics in order to be representative. In principle, the material must not exhibit macroscopic heterogeneity. This is not always the case for DP steels. Many DP steels contain pronounced banding structures along the rolling direction in their cross section. However, this effect is commonly ignored in the literature (at least in the cited references of this work). Considering DP steels, the size of the volume element must be large enough to contain a statistically significant number of ferrite and martensite grains. The calculated material behavior must not change for different volume-element sizes. It is remarked that the geometrical parameters in this work are either based on the largest available two-dimensional EBSD scan or are directly taken from the data in the source articles.

1.4 Scale-bridging techniques

A multitude of different scale-bridging techniques exists in the literature. The commonly shared goal is the efficient incorporation of the governing microstructural behavior to determine the macroscopic material response. This can be approached directly by the incorporation of microstructural parameters into macroscopic constitutive-laws, e.g., the Hall-Petch grain-size effect or the scaling of material parameters in more elaborate monolithic material models (e.g., Resende and Bouvier, 2013).

One popular class of scale-bridging techniques makes use of numerical RVE calculations. The dedicated and physically-detailed full-field simulations can be carried out with the finite-element method, fast-Fourier transformation or other applicable solution techniques. With the numerical RVE results, several different approaches can be considered. The identification of material parameters for macroscopic material models (virtual material testing). Or using of the RVE results as direct input for finite-element simulations at the microscale or Gauss-point level (FE^2 method). Finally, the RVE results might be utilized as input for the identification of microscopic strain fields that can be piecewise uniform (Transformational field analysis, TFA) or nonuniform (NTFA) (Michel and Suquet, 2003).

The scale-bridging approach adopted for this work, belongs to the group of analytical or semi-analytical methods that has originally been developed for linear problems. If solely the volume fraction is taken into account, simple analytical methods give rigorous first-order bounds for elastic materials. Voigt's estimation assumes uniform strain on the micro level and, thus, trivially fulfills kinematic compatibility, but violates static compatibility. The resulting effective material-stiffness is the arithmetic mean of the single phase stiffnesses and, thus, the upper bound for the material stiffness. In the nonlinear case, equal strain on the micro level (iso-strain) is commonly referred to as Taylor assumption. Reuss' estimation assumes equal stresses in all phases and, thus, violates kinematic compatibility. The resulting effective elastic stiffness is the harmonic mean of the respective phase stiffnesses and constitutes the lower bound for the material stiffness. In the nonlinear case, this bound, based on equal stresses (iso-stress), is known as Sachs assumption (Gross and Seelig, 2011). There are several other analytical approaches, e.g., the equi-incremental mechanical work approach of Bouaziz and Buessler (2002) not necessitating any additional fitting parameters.

If the phase properties vary significantly, the two first order bounds yield a large spectrum of possible effective properties. In order to approximate fluctuations and interactions in more detail, refined estimation approaches rely on the analytical Eshelby solution for ellipsoidal inclusions. Classic examples are, e.g., the dilute-distribution model, the Mori-Tanaka model (which coincides with the lower second-order Hashin-Shtrikman bound for the isotropic case), the differential-scheme and self-consistent estimates (Gross and Seelig, 2011). Closer bounds can be obtained by the variational approach of Hashin and

Shtrikman (Willis, 1977). Originally developed for linear material behavior, many of the mentioned (semi-)analytical techniques have been applied and modified for non-linear material behavior. Because a Hashin-Shtrikman based approach is used in the work at hand, the fundamentals of this approach is covered in more detail in Chapter 2.

1.5 Influence of grain size distributions

Polycrystalline metals like DP steels are often reported to have logarithmically-normal distributed grain diameters (Valiev et al., 2000; Nicaise et al., 2011). However, most constitutive material models, e.g., the Hall-Petch relation, only consider the mean grain-diameter and, thus ignore the statistical distribution of grain sizes.

Relatively few contributions account for grain-size distribution functions, e.g., Kurzydlowski (1990), Masumura et al. (1998). Within both works, the polycrystalline behavior is modeled assuming homogeneous strains (iso-strain model). In recent years, the problem of a physically motivated relation between macroscopic flow-behavior and grain-size distribution has gradually received more attention.

Berbenni et al. (2007) observe a decreasing yield-strength with increasing grain-diameter dispersion; they employ a self-consistent, elasto-viscoplastic homogenization scheme and an ideal-plastic microscopic material-behavior. Internal stresses and the stored energy due to internal stresses rise with increased grain-size dispersion.

Raeisinia (2008) and Raeisinia et al. (2008) extend the local modeling, utilizing a more sophisticated microscopic material-model and incorporating a grain-size dependent initial yield-strength and hardening. A simple statistical iso-strain model for the elasto-plastic transition of polycrystals can be found in Raeisinia and Poole (2012).

Focusing on different deformation mechanisms on the grain level, Zhu et al. (2005, 2006) investigated nanocrystalline and face-centered cubic aggregates. In the work of Ramtani et al. (2009), nanocrystalline copper is modeled as a composite material with a log-normal distributed grain-size. Nicaise et al. (2011) improve the complexity by investigating coupled effects of grain size distributions and crystallographic textures for interstitial-free steels.

Most of the articles cited above, highlight the influence of a log-normal grain-size distribution on the macroscopical behavior, especially for small average grain-diameters. To the best knowledge of the author, an experimental examination of grain-size distributions is generally challenging. Whenever either of the two most important production-parameters (either the alloy composition or thermo-mechanical treatment) is altered, both the microstructure and mechanical behavior at grain level change simultaneously. One cannot create two different microstructures where only the microstructure *or* the mechanical behavior at the grain level changes.

1.6 Outline

The experimental methods that have been applied to a DP600 steel to obtain a data basis for this work are summarized in Chapter 5. Chapter 2 summarizes the nonlinear mean-field model that is employed to define the strain localization from the finite-element integration points to the constituents of the DP steel. In Chapter 3, the influence of grain-size distributions on mean-field modeling is studied. The work-hardening of DP steels is investigated on the grain-level and in a statistical manner in Chapter 4. Chapter 6 explains the physical motivation of the proposed material-model for DP steels. Simulation results are discussed in Chapter 7, and the influence of different microstructural parameters is investigated. Chapter 8 concludes this work.

1.7 Notation

A direct tensor notation is preferred throughout the text. For tensor components, Latin indices are used and Einstein's summation convention is applied. Vectors and second-order tensors are denoted by lowercase and uppercase bold letters. The composition of two second-order or two fourth-order tensors is formulated by, e.g., $\boldsymbol{AB}$ and $\mathbb{AB}$, respectively. A linear mapping of second-order tensors by a fourth-order tensor is written as $\boldsymbol{A} = \mathbb{C}[\boldsymbol{B}]$. Scalar and dyadic products are denoted, e.g., by $\boldsymbol{A} \cdot \boldsymbol{B}$ and $\boldsymbol{A} \otimes \boldsymbol{B}$, respectively. The second-order identity is denoted by $\boldsymbol{I}$, and the identity on symmetric second-order tensors by $\mathbb{I}^{\mathrm{S}}$. Completely symmetric and traceless tensors are designated by a prime, $\boldsymbol{A}'$, while spherical tensors are denoted by a circle, $\boldsymbol{A}^{\circ}$. The spherical projector is given by $\mathbb{P}_{\mathrm{sph}} = \boldsymbol{I} \otimes \boldsymbol{I}/3$ and the deviatoric projector by $\mathbb{P}_{\mathrm{dev}} = \mathbb{I}^{\mathrm{S}} - \mathbb{P}_1$. The symmetric part of a second-order tensor is denoted by $\mathrm{sym}\,(\boldsymbol{A})$, and the skew-symmetric part by $\mathrm{skw}\,(\boldsymbol{A})$. DP$x$ denotes a dual-phase steel with an ultimate tensile strength of x, given in MPa.

1.8 Preliminaries

This section provides general definitions for. It follows the more complete description of the continuum theory in Bertram (2008).

Kinematics The position of each material point of a body can be represented by means of the position vector in either the reference configuration $\hat{\boldsymbol{X}}$, or in the current configuration $\boldsymbol{x}$. Using the deformation χ, the current configuration is given as an invertible function of the reference configuration at any time t by

$$\boldsymbol{x} = \chi\left(\hat{\boldsymbol{X}}, t\right). \tag{1.2}$$

The displacement vector $\boldsymbol{u}$ denotes the difference in the material position at all times,

$$\boldsymbol{u}\left(\hat{\boldsymbol{X}}, t\right) = \chi\left(\hat{\boldsymbol{X}}, t\right) - \hat{\boldsymbol{X}}. \tag{1.3}$$

The partial derivative of Equation (1.3) with respect to the reference configuration $\hat{\boldsymbol{X}}$ defines the displacement gradient $\boldsymbol{H}$ and, implicitly, the deformation gradient $\boldsymbol{F} = \mathrm{Grad}\,(\boldsymbol{x})$,

$$\boldsymbol{H} = \mathrm{Grad}\,(\boldsymbol{u}) = \frac{\partial \boldsymbol{u}\left(\hat{\boldsymbol{X}}, t\right)}{\partial \hat{\boldsymbol{X}}} = \frac{\partial\left(\boldsymbol{x}\left(\hat{\boldsymbol{X}}, t\right) - \hat{\boldsymbol{X}}\right)}{\partial \hat{\boldsymbol{X}}} = \boldsymbol{F} - \boldsymbol{I}. \tag{1.4}$$

Using the right Cauchy-Green tensor $\boldsymbol{C} = \boldsymbol{F}^{\mathsf{T}}\boldsymbol{F}$, Green's strain-tensor and its representation by the displacement gradient is given by

$$\boldsymbol{E} = \frac{1}{2}\left(\boldsymbol{C} - \boldsymbol{I}\right) = \frac{1}{2}\left(\boldsymbol{F}^{\mathsf{T}}\boldsymbol{F} - \boldsymbol{I}\right) = \frac{1}{2}\left(\boldsymbol{H} + \boldsymbol{H}^{\mathsf{T}} + \boldsymbol{H}^{\mathsf{T}}\boldsymbol{H}\right). \tag{1.5}$$

For small deformations, $\|\boldsymbol{H}\| \ll 1$, one finds the infinitesimal strain-tensor $\boldsymbol{\varepsilon}$ as linearization of Green's strain-tensor (and other generalized finite strain quantities) by

$$\boldsymbol{\varepsilon} = \frac{1}{2}\left(\boldsymbol{H} + \boldsymbol{H}^{\mathsf{T}}\right) = \mathrm{sym}\,(\boldsymbol{H}). \tag{1.6}$$

Infinitesimal strain is complementary to infinitesimal rotation $\boldsymbol{\omega} = \mathrm{skw}\,(\boldsymbol{H})$. The infinitesimal strain-tensor (and all other second-order tensors) can be additively decomposed into a spherical part ε° and a deviatoric part ε' in a unique way by

$$\varepsilon^{\circ} = \frac{1}{3}\mathrm{sp}\,(\varepsilon)\,\boldsymbol{I}, \qquad \varepsilon' = \varepsilon - \varepsilon^{\circ}. \tag{1.7}$$

In the case of infinitesimal strains, the spherical part is equivalent to the volumetric change of the body (dilatation) and the deviatoric part gives the shape-altering deformation for a constant volume (isochoric). The isotropic material-models of Chapter 3, Section 4.2.2 and Chapter 6 make use of the assumption of an additive decomposition of the infinitesimal strain into its elastic ε^{e} and plastic ε^{p} part, as well as a thermal strain ε^{θ},

$$\varepsilon = \varepsilon^{\mathrm{e}} + \varepsilon^{\mathrm{p}} + \varepsilon^{\theta}. \tag{1.8}$$

The thermal strain is assumed to be of purely dilatational structure, i.e., it is given by the expression $\varepsilon^{\theta} = \alpha\boldsymbol{I}$, here.

Constitutive equations For the mean-field material models in this work, the inner stress of a material (Cauchy stress-tensor $\boldsymbol{\sigma}$) is assumed to obey Hooke's law in the form

$$\boldsymbol{\sigma} = \mathbb{C}[\varepsilon - \varepsilon^{\mathrm{p}} - \varepsilon^{\theta}]. \tag{1.9}$$

Here, $\mathbb{C}$ is the stiffness-tensor of the material. The (hyper) elastic behavior is assumed to be isotropic, i.e., $\mathbb{C} = 3K\mathbb{P}_1 + 2G\mathbb{P}_2$. The material parameter K denotes the compression modulus and G denotes the shear modulus of the material. Rate-independent and associated plasticity is assumed, thus the evolution of the plastic strain ε^{p} is obtained from a potential relation via

$$\dot{\varepsilon}^{\mathrm{p}} = \dot{\gamma}\,\frac{\partial\varphi(\boldsymbol{\sigma}', \boldsymbol{X}, \sigma^{\mathrm{f}})}{\partial\boldsymbol{\sigma}'}, \tag{1.10}$$

of the von Mises yield-function

$$\varphi(\boldsymbol{\sigma}', \boldsymbol{X}, \sigma^{\mathrm{f}}) = \|\boldsymbol{\sigma}' - \boldsymbol{X}\| - \sqrt{\frac{2}{3}}\sigma^{\mathrm{f}}\left(\varepsilon^{\mathrm{p}}\right) = 0. \tag{1.11}$$

The accumulated plastic-strain,

$$\varepsilon^{\mathrm{p}} = \sqrt{\tfrac{2}{3}} \int_0^t \|\dot{\boldsymbol{\varepsilon}}^{\mathrm{p}}\| \, \mathrm{d}\bar{t}, \tag{1.12}$$

and the kinematic stress-tensor $\boldsymbol{X}$ (or back-stress tensor, see, e.g., Kassner et al. (2012) for a review of this topic) are the internal variables. Their evolution equations are discretized with an implicit Euler time-integration scheme in this work. See, e.g., Simo and Hughes (1998) for a more profound discussion of the von-Mises model. The evolution of the kinematic stress follows an expression equivalent to the Armstrong-Frederick relation (Frederick and Armstrong, 2007; Resende and Bouvier, 2013)

$$\mathrm{d}\boldsymbol{X} = c\left(X_{\mathrm{sat}}\left(\rho, \mathrm{MS}\right)\boldsymbol{N} - \boldsymbol{X}\right)\mathrm{d}\varepsilon^{\mathrm{p}}. \tag{1.13}$$

The evolution of the kinematic saturation-stress $X_{\mathrm{sat}}\left(\rho, \mathrm{MS}\right)$ is part of the proposed material model and is derived in Chapter 6.

Chapter 2

Mean-field model

2.1 Hashin-Shtrikman based mean-field model

In this work, different material models of constituents or phases interact through a non-linear mean-field model[1]. "Phase" is used in the sense of a domain with identical material properties. The employed mean-field model is a non-linear extension of the Hashin-Shtrikman scheme, summarized below from the articles Böhlke et al. (2014) and Rieger and Böhlke (2015). For a more detailed description, the reader is referred to the original works of Jöchen and Böhlke (2012) and Jöchen (2013).

Homogenization based on piecewise constant-stress polarizations. The effective material behavior is modeled based on a linear-elastic homogeneous comparison-material (Willis, 1977; Ponte Castaneda et al., 1997). Stress polarizations are given, relative to a comparison medium, with the stiffness tensor $\mathbb{C}_0$,

$$\boldsymbol{p}(\boldsymbol{x}) = \boldsymbol{\sigma}(\boldsymbol{x}) - \mathbb{C}_0[\boldsymbol{\varepsilon}(\boldsymbol{x})]. \tag{2.1}$$

[1] This section is based on Böhlke et al. (2014) and Rieger and Böhlke (2015)

The stress tensor is then decomposed by

$$\boldsymbol{\sigma}(\boldsymbol{x}) = \mathbb{C}_0[\boldsymbol{\varepsilon}(\boldsymbol{x})] + \boldsymbol{p}(\boldsymbol{x}), \tag{2.2}$$

and, thus, a modified boundary value problem is given by

$$\operatorname{div}\left(\mathbb{C}_0[\boldsymbol{\varepsilon}(\boldsymbol{x})]\right) + \operatorname{div}\left(\boldsymbol{p}(\boldsymbol{x})\right) = \mathbf{0}. \tag{2.3}$$

The polarization $\boldsymbol{p}(\boldsymbol{x})$ of an RVE describes the difference between the actual (heterogeneous) stresses $\boldsymbol{\sigma}(\boldsymbol{x})$ and the stress $\mathbb{C}_0[\boldsymbol{\varepsilon}(\boldsymbol{x})]$ of a homogeneous comparison RVE with identical local strain. Usage of the stress polarization instead of the actual stresses, however, limits the influence of approximation errors on the overall result (Gross and Seelig, 2011). In the following, piecewise constant trial polarizations are assumed in each phase, i.e.,

$$\boldsymbol{p}(\boldsymbol{x}) = \sum_{\alpha=1}^{N} \chi_\alpha(\boldsymbol{x})\boldsymbol{p}_\alpha, \tag{2.4}$$

where $\chi_\alpha(\boldsymbol{x})$ denotes the indicator function of phase α. The polarizations $\boldsymbol{p}(\boldsymbol{x})$ are assumed to be known. Then, strains that solve the boundary-value problem in Equation (2.3) can be simplified in the sense that they are given by the ensemble average

$$\boldsymbol{\varepsilon}_\alpha = \bar{\boldsymbol{\varepsilon}} + \frac{1}{c_\beta}\sum_{\beta=1}^{N} \mathbb{G}_{\alpha\beta}[\boldsymbol{p}_\beta]. \tag{2.5}$$

Here, $\bar{\boldsymbol{\varepsilon}}$ is the effective strain tensor, $\boldsymbol{\varepsilon}_\alpha = \langle\boldsymbol{\varepsilon}\rangle_\alpha$ are the average phase strains, the volume fractions of the domains are denoted by c_α and N is the number of phases with $\alpha = 1 \ldots N$. The fourth-order tensor $\mathbb{G}_{\alpha\beta}$ includes the second derivatives of Green's function for an infinite body (Ponte Castaneda et al., 1997) and depends on the comparison material and the microstructure of the actual material (Jöchen, 2013).

Implications of statistical isotropy and no long-range order. Under the hypotheses of statistical homogeneity and no long-range order, Willis (1977) has shown that the micro-structural tensor is given by $\mathbb{G}_{\alpha\beta} = c_\alpha(\delta_{\alpha\beta} - c_\beta)\mathbb{P}_0$, with the polarization tensor $\mathbb{P}_0$. Without loss of generality, the comparison material can be chosen to be isotropic (Nadeau and Ferrari, 2001). For an isotropic comparison-medium with a stiffness $\mathbb{C}_0 = 3K_0\mathbb{P}_{\text{sph}} + 2G_0\mathbb{P}_{\text{dev}}$ and isotropic two-point statistics, the polarization tensor $\mathbb{P}_0$ is given by

$$\mathbb{P}_0 = p_1\mathbb{P}_{\text{sph}} + p_2\mathbb{P}_{\text{dev}}. \tag{2.6}$$

Here, $p_1 = (3K_0 + 4G_0)^{-1}$ and $p_2 = 3(K_0 + 2G_0) / (5G_0(3K_0 + 4G_0))$ (Dederichs and Zeller, 1973).

Bounds and $\mathbb{C}_0$ based estimates. With Equation (2.5) and the expression for the polarization tensor $\mathbb{P}_0$, the following equations are obtained for each constituent

$$\boldsymbol{\sigma}_\alpha - \langle\boldsymbol{\sigma}\rangle = \mathbb{L}[\boldsymbol{\varepsilon}_\alpha - \langle\boldsymbol{\varepsilon}\rangle]. \tag{2.7}$$

Here, $\boldsymbol{\sigma}_\alpha = \langle\boldsymbol{\sigma}\rangle_\alpha$ denotes the phase average of the stress and $\boldsymbol{\varepsilon}_\alpha = \langle\boldsymbol{\varepsilon}\rangle_\alpha$ the respective average of strains. The Hill tensor is given by $\mathbb{L} = \mathbb{C}_0 - \mathbb{P}_0^{-1}$. Due to the fact that $||\mathbb{L}|| \to 0$ if $||\mathbb{C}_0|| \to 0$ and $||\mathbb{L}|| \to \infty$ if $||\mathbb{C}_0|| \to \infty$, the special cases $||\mathbb{C}_0|| \to 0$ and $||\mathbb{C}_0|| \to \infty$ yield the simple bounds by Reuss (homogeneous stresses $\boldsymbol{\sigma}_\alpha = \bar{\boldsymbol{\sigma}}$) and by Voigt (homogeneous strains $\boldsymbol{\varepsilon}_\alpha = \bar{\boldsymbol{\varepsilon}}$) (Ponte Castaneda et al., 1997). These first-order bounds are obtained for both linear and nonlinear constitutive behavior of the domains. For a linear elastic or linear thermo-elastic material, the set of equations given by Equation (2.5) can be solved explicitly. The well known Hashin-Shtrikman bounds are obtained as special cases for the specific choices of the comparison materials. If the stiffness tensor of the (homogeneous) comparison material $\mathbb{C}_0$ is chosen to satisfy $\mathbb{C} - \mathbb{C}_0 \leq 0$ for a quadratic form in $\langle \varepsilon \rangle$, the minimum comparison stiffness results in an upper bound of the strain-energy density. If $\mathbb{C} - \mathbb{C}_0 \geq 0$ is chosen, the maximum comparison stiffness results in a lower bound. This lower bound is also exactly obtained from the Mori-Tanaka model in the isotropic case. In case of nonlinear constitutive behavior, it is noted, that the aforementioned Hashin-Shtrikman bounds are not rigorous.

Governing equations. The material description is used for infinitesimal strains only. To allow at least for an approximation of the finite-strain behavior in the finite-element calculations, the material model is now formulated in incremental form. The incremental form of Hooke's law for a single phase α is given by

$$\Delta \boldsymbol{\sigma}_\alpha = \mathbb{C}_\alpha \left[\Delta \boldsymbol{\varepsilon}_\alpha - \Delta \boldsymbol{\varepsilon}^{\mathrm{p}}_\alpha \right]. \tag{2.8}$$

The local phase strain can be deduced from the macroscopic strain increment $\Delta \bar{\boldsymbol{\varepsilon}}$ by employing the mean-field model as described above,

$$\Delta \boldsymbol{\varepsilon}_\alpha = \mathbb{A}^\varepsilon \left[\Delta \bar{\boldsymbol{\varepsilon}} \right] - \mathbb{R} \left\langle \mathbb{C} \left[\Delta \boldsymbol{\varepsilon}^{\mathrm{p}} \right] \right\rangle + \mathbb{R} \mathbb{C}_\alpha \left[\Delta \boldsymbol{\varepsilon}^{\mathrm{p}} \right]. \tag{2.9}$$

The fourth-order strain localization tensor $\mathbb{A}^\varepsilon$ reads

$$\mathbb{A}^\varepsilon = \mathbb{R} \left\langle \mathbb{R} \right\rangle^{-1}, \tag{2.10}$$

and is given by means of the microstructural tensor

$$\mathbb{R} = \left(\mathbb{P}_0^{-1} - (\mathbb{C} - \mathbb{C}_0) \right)^{-1}. \tag{2.11}$$

2.2 Simplification for identical isotropic elastic-properties

For the special case, where $(\mathbb{C}_\alpha - \mathbb{C}_0) \propto \mathbb{P}_{\text{dev}}$ and $\mathbb{C}_\alpha = \mathbb{C} = 3K\mathbb{P}_{\text{sph}} + 2G\mathbb{P}_{\text{dev}}$, the governing Equations (Equation (2.8), Equation (2.9)) can be simplified considerably (Glavas, 2014). The former relation, $(\mathbb{C}_\alpha - \mathbb{C}_0) \propto \mathbb{P}_{\text{dev}}$, implies that $\mathbb{C}_\alpha$ and $\mathbb{C}$ only differ in their deviatoric part. This is equivalent to different shear moduli G among the constituents while the compression moduli K are identical for the comparison material $\mathbb{C}_0$ and the constituents. This assumption reflects that hydrostatic pressure is often considered secondary in metals; it does not significantly alter the deformation behavior or yield stress of most metals (e.g., Gambin, 2002). Consequently, plastic deformation is often assumed to be isochoric; an assumption the applied von Mises model uses, too. In this work, the mean-field model is mainly employed for the distribution of plastic eigenstrains which remains unaltered by the former assumption. The second relation, $\mathbb{C}_\alpha = \mathbb{C}$, assumes that ferrite and martensite have the same shear moduli, too. Experimentally observed differences of the shear moduli would alter the DP-simulation results only slightly and mainly in the initial elastic part, during which both phases still behave elastic. As soon as one phase starts to yield, comparison simulations show that the work-hardening modulus is the main influence on the overall behavior and slight differences in the shear-modulus are negligible.

From these assumptions, it follows that the microstructural tensor $\mathbb{R}$ in Equation (2.10) is constant and homogeneous for all phases which in turn yields the fourth order symmetric unity tensor

$$\mathbb{A}^\varepsilon = \mathbb{I}^{\text{S}}. \tag{2.12}$$

For the microstructural tensor one then obtains

$$\mathbb{R} = p_1 \mathbb{P}_{\text{sph}} + r_2 \mathbb{P}_{\text{dev}}, \tag{2.13}$$

with p_1 and p_2 from the definition of $\mathbb{P}_0$ in Equation (2.6) and $r_2 = \left(p_2^{-1} + 2\left(G - G_0\right)\right)^{-1}$. Introducing the two results, the governing equations (Equation (2.8) and Equation (2.9)) reduce to

$$\Delta\boldsymbol{\sigma}_\alpha = \mathbb{C}\left[\Delta\boldsymbol{\varepsilon}_\alpha - \Delta\boldsymbol{\varepsilon}^{\mathrm{p}}_\alpha\right], \tag{2.14}$$

$$\Delta\boldsymbol{\varepsilon}_\alpha = \Delta\bar{\boldsymbol{\varepsilon}} - \lambda\left\langle\Delta\boldsymbol{\varepsilon}^{\mathrm{p}}_\alpha\right\rangle + \lambda\Delta\boldsymbol{\varepsilon}^{\mathrm{p}}_\alpha, \tag{2.15}$$

where $\lambda \in [0, 1]$ is a single scalar parameter, which reads

$$\lambda\left(G, G_0, K_0\right) = \frac{2G}{\left(p_2^{-1} + 2\left(G - G_0\right)\right)}. \tag{2.16}$$

Assuming identical compression-moduli for all constituents, the parameter λ is only dependent on the shear-modulus difference between comparison material and constituents.

The comparison material, or in the simplified case the parameter λ, governs the fluctuations of the stress and strain fields and therefore the behavior of the mean-field model. For all choices of λ, physically admissible stress and strain estimates are obtained. If $\lambda = 0$, the strain fluctuations vanish and the Voigt-bound is recovered by $\Delta\boldsymbol{\sigma}_\alpha = \mathbb{C}\left[\Delta\bar{\boldsymbol{\varepsilon}} - \Delta\boldsymbol{\varepsilon}^{\mathrm{p}}_\alpha\right]$. On the opposite, if $\lambda = 1$, the stress fluctuations vanish and the Reuss-bound is recovered by $\Delta\boldsymbol{\sigma}_\alpha = \mathbb{C}\left[\Delta\bar{\boldsymbol{\varepsilon}} - \left\langle\boldsymbol{\varepsilon}^{\mathrm{p}}_\alpha\right\rangle\right] = \Delta\bar{\boldsymbol{\sigma}}$. The actual choice of λ is dictated by the fact that both the macroscopic behavior and strain-partitioning on the microscopic level must correspond to experimental data. In this sense, λ cannot be chosen freely to fit the model. It is imposed by the experimentally measured macro- and microscopic stress-strain partitioning.

2.3 Numerical implementation

For nonlinear elastic materials, stress is a phase-dependent nonlinear function of strain. In the inelastic case, after introducing a time-integration scheme for the internal variables, the stress is also given by a generally nonlinear function of strain. In most inelastic cases this function will be implicit. The integration of internal variables extends the set of equations determining the stress by the respective evolution equations. Thus, iterative solution schemes have to be applied for nonlinear constitutive models. To calculate the localization within the mean-field model, a staggered iteration-scheme is applied for the proposed material model. At first, the ferrite material state is calculated. Then, the long-range stress tensor from the ferrite model is passed on to the martensite material model and treated as a constant input value.

For the proposed material model (Chapter 6), time integration is challenging since both phases heavily interact with each other through long-range stresses that evolve with deformation. As will be described in Chapter 6, there is a significant non-linearity within the definition of the long-range stresses. Additionally, the evolution of the long-range stresses is much faster than the evolution of the regular work-hardening. Therefore, the number of iterations until the mean-field model converges depends on the time increments: $50 - 100$ time increments were used for the simulations in this work. In Chapter 3 and Chapter 4, non-interacting material models are employed. For these models, the number of necessary localization iterations for a given finite element time-increment increases with the number of phases.

Chapter 3

Influence of grain-size distribution on mean-field modeling

Metals, such as DP steels, are often reported to have logarithmically-normal grain-diameter distributions (Valiev et al., 2000; Nicaise et al., 2011). Therefore, the influence of micro-heterogeneous deformation on the grain-level is investigated in this section[1]. Due to its computational efficiency, the mean-field method that is presented in Chapter 2 is well suited to investigate statistical influences.

3.1 Microstructures

Two types of RVEs are used for the simulations. For Section 3.4.1, the used analytical RVEs are based on the assumption of spatially randomly distributed spherical grains that obey a log-normal grain-diameter distribution. The experimental RVE considered in Section 3.5 is obtained by evaluating two-dimensional EBSD data.

With the average grain-diameter μ and the standard deviation σ_{μ}, a log-normal distribution can be defined by two parameters

[1] This section is reproduced from the author´s contribution to Böhlke et al. (2014).

$S = \sqrt{\ln\left(\sigma_{\mu}/\mu^2 + 1\right)}$ and $M = \ln(\mu) - S^2/2$. With the two defined parameter one yields

$$P_{\mathrm{Dia.}}(D) = \frac{1}{\sqrt{2\pi}SD} \exp\left(-\frac{(\ln(D) - M)^2}{2S^2}\right), \tag{3.1}$$

where D denotes the grain-size. For modeling analytical RVEs, the volume-probability function can be calculated from Equation (3.1),

$$P_{\mathrm{Vol}}(D) = \frac{1}{\sqrt{2\pi}SD} \exp\left(-\frac{\left(\ln(D) - \left(M + 3S^2\right)\right)^2}{2S^2}\right). \tag{3.2}$$

For this three-dimensional case the identity $\int_0^\infty D^q P_{\mathrm{Dia.}} \mathrm{d}D = \exp\left(qM + q^2S^2/2\right)$ holds (Raeisinia, 2008). Note that the volume distribution is still of log-normal type, and that only M changes to be $M+3S^2$ in Equation (3.2).

The discrete three-dimensional diameter distribution for an experimental RVE cannot be calculated exactly from a two-dimensional EBSD scan. Therefore, the area-probability function must be calculated as well. With a calculation analogous to Raeisinia (2008) the area distribution for comparison with the EBSD based RVE is given by the similar expression

$$P_{\mathrm{Area}}(D) = \frac{1}{\sqrt{2\pi}SD} \exp\left(-\frac{\left(\ln(D) - \left(M + 2S^2\right)\right)^2}{2S^2}\right). \tag{3.3}$$

It is remarked that grain diameters obtained from two-dimensional cuts constitute only a lower bound to the actual grain diameters. Nevertheless, it is possible to calculate statistic moments for the corresponding three-dimensional grain-size distribution ("stereological unfolding-problem"). Assuming a log-normal experimental grain-size

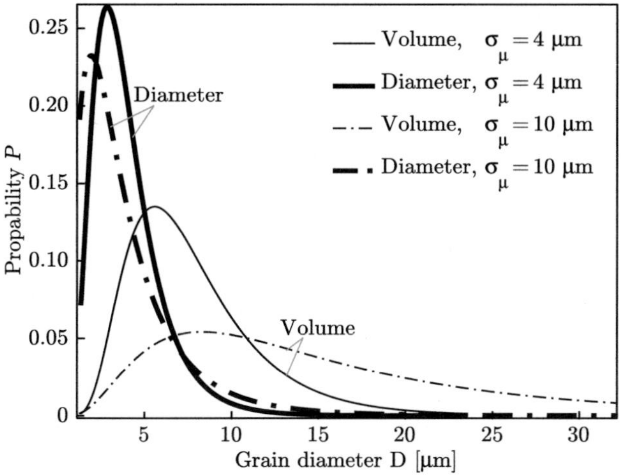

Figure 3.1: Analytical grain diameter and volume distributions for a single mean diameter $\mu = 4\,\mu\mathrm{m}$ and two different standard deviations σ_{μ}. Note the opposite trend for volume and diameter distributions with increasing standard deviation.

distribution that is defined in terms of the first two area moments (μ, σ_{μ}) of the EBSD scan, the three-dimensional distribution can be estimated from the two-dimensional data (e.g., Ohser and Mücklich, 2000). The high efforts to determine slightly better diameter distributions, however, has led to a broad application of simple two-dimensional grain-size and area distributions. Alternatively, only an average diameter is used for micromechanics. It is emphasized that different definitions of the average diameter with different micromechanical implications are possible.

In order to discretize the analytical area and volume distributions for the mean-field model, a discretization range is defined that accounts for at least $\int_{D_{\min}}^{D_{\max}} P_{\mathrm{Vol.}}(D)\,\mathrm{d}D = 0.98$ of the probability-density distribution. The discretization range is then subdivided into equally distributed grain diameter intervals. Each interval is represented by its

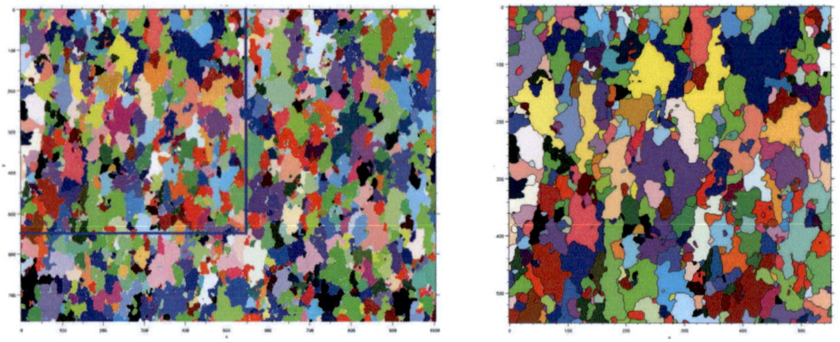

(a) (b)

Figure 3.2: Original EBSD data for DC04 steel (Schreijäg, 2013) (a); identified grains (b). The blue square indicates the (equivalent) reduced data set for simulations in this work.

average grain-diameter and a constant probability equal to the total probability of the particular interval. Numerical simulations show that already fifty grain-size intervals are sufficient to model the effects observed for analytical grain-size distributions in this work.

A low-alloyed single-phase ferritic DC04 steel is considered in order to apply the investigation results to an experimental grain-size distribution. The microstructure has been characterized by an EBSD scan (see Section 5.2 for details about the experimental setup). Subsequently, the EBSD data has been analyzed using the Matlab MTEX toolbox (Bachmann et al., 2011). Grains have been identified using a segregation angle of 15°. Figure 3.2(a) shows the full original EBSD data and Figure 3.2(b) depicts the 637 identified grains. In order to reduce the amount of grains for the simulations, a smaller area of the complete EBSD scan (blue square in Figure 3.2(a)) has been chosen as input for the RVE simulations, referenced as "reduced" data set in the following.

The microstructural input for the simulation consists of two interdependent variables that are calculated for each RVE grain. At first, the

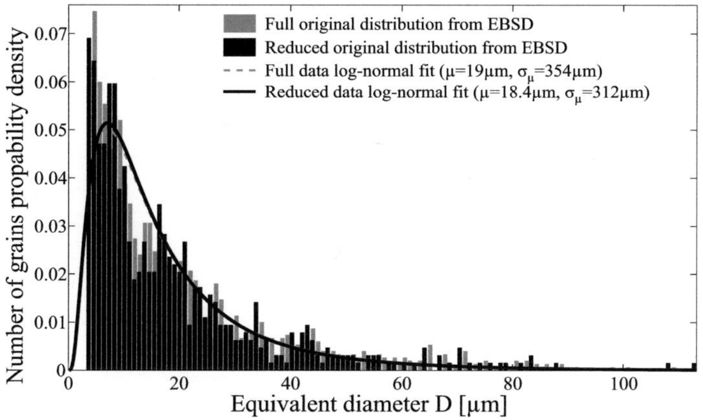

Figure 3.3: Grain diameter distributions for the full and reduced EBSD data-sets and both fitted analytical log-normal distributions. The fitted parameters are in good agreement despite the data reduction, however, there is a gap between $10 - 20\ \mu\mathrm{m}$.

phase fraction c_α for the homogenization scheme is calculated using the respective area fraction from the EBSD scan. Then, the equivalent diameter of each grain is calculated by assuming a circular grain with the same area as the original grain.

Figure 3.3 shows the grain-diameter distribution for the full and reduced EBSD data as well as log-normal fits for both data sets. The log-normal fits exhibit good overall agreement. However, the diameter distribution shows a concentration of grains in the region up to $10\ \mu\mathrm{m}$. One can also observe that there is an underproportional amount of grains the region of $10 - 20\ \mu\mathrm{m}$.

3.2 Constitutive equations

Focusing on a basic understanding of the impact of grain-size distributions on the mechanical response, the material models for this inves-

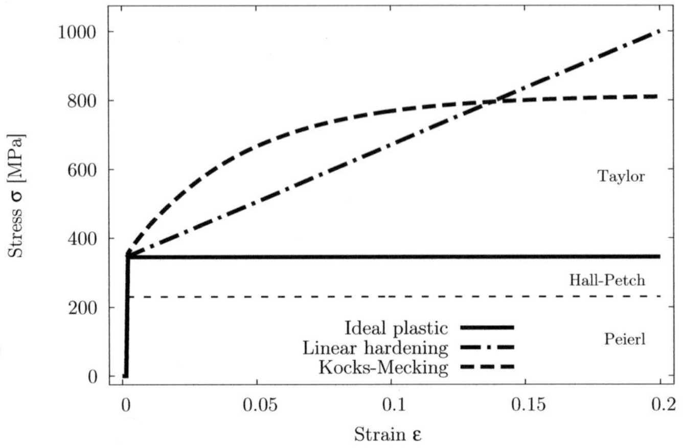

Figure 3.4: Uniaxial tension stress-strain diagrams for the three material models used to investigate the effect of grain-size distributions

tigation are rather simple (see Section 6.1 for a more extensive model description). The more complex model that is proposed later in this work is, however, not used for the investigation in this chapter. Albeit being phenomenological and macroscopic, it is assumed that the employed constitutive relations hold for single grains, too.

The first two models employ a Hall-Petch relation for the initial yield-stress and assume either no hardening (ideal-plastic model) or size-independent and linear-isotropic hardening (linear-hardening model). The third model is a standard Kocks-Mecking model for monotonic loading (Kocks and Mecking, 2003). The employed relation for the yield stress comprises two contributions,

$$\sigma^{\mathrm{f}} = \underbrace{\sigma_0 + \frac{k_{\mathrm{HP}}}{\sqrt{D}}}_{\text{Hall-Petch}} + \underbrace{MG\alpha b\sqrt{\rho}}_{\text{Taylor}} . \tag{3.4}$$

Parameter		Value
Elastic modulus	E	220 GPa
Poisson´s ratio	ν	0.3
Dislocation interaction	α	0.3
Magnitude of burgers vector	b	0.25 nm
Taylor factor	M	3.06
Initial dislocation density	ρ_0	$10^{11}\,\mathrm{m}^{-2}$
Dislocation storage	k_1	$0.09\,\mathrm{nm}^{-1}$
Dislocation recovery	k_2	13.68
Yield stress constant	σ_0	77 MPa (230 MPa)
Hall-Petch coefficient	k_{HP}	$16\,\mathrm{MPa}\sqrt{\mathrm{mm}}$ ($150\,\mathrm{MPa}\sqrt{\mathrm{m}}$)

Table 3.1: Material parameters for the simulations of Section 3.5 and Section 3.4.1 (enclosed in brackets). The values for Section 3.4.1 are artificial, based on the values for DP steel in Delince et al. (2007).

The constant initial yield-strength follows from the Hall-Petch relation, while the isotropic Taylor-type hardening is grain-size independent. The total (average) dislocation-density ρ is used as an internal variable to account for the forest hardening of ferrite. The evolution equation for the total dislocation density

$$\mathrm{d}\rho = M\left(k_{\mathrm{f}}\sqrt{\rho} - k_{\mathrm{a}}\rho\right)\mathrm{d}\varepsilon^{\mathrm{p}} \tag{3.5}$$

consists of an athermal storage, $k_{\mathrm{f}}\sqrt{\rho}$, that is related to the mean free path of dislocations $\sqrt{\rho}$, and a recovery term, $-k_{\mathrm{a}}\rho$, that is proportional to the current dislocation density ρ. No kinematic hardening is taken into account for this investigation on grain-size distributions.

3.3 Material parameters

Micro-pillar compression experiments from the work of Schreijäg (2013) are used to identify the microscopical material parameters (Figure 3.5). A low elastic slope of micro-pillar compression measurements

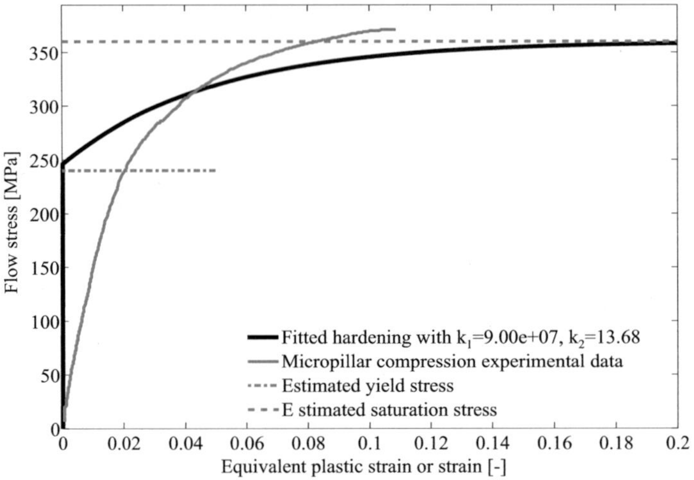

Figure 3.5: Experimental micropillar compression data from Schreijäg (2013) compared to the chosen ferrite hardening.

is a common phenomenon that may be attributed to localized plasticity or stress and deformation inhomogeneities near the flat punch. Work-hardening of each grain after initial yielding is assumed to be independent of the individual grain size. Therefore, micro-pillar compression results are solely used to define the material hardening parameters k_1/k_2. The modeled hardening-stress magnitude for the saturated dislocation-density on the grain level, is concurrent with the experimentally observed microscopic hardening. All model parameters are summarized for reference in Table 3.1.

The hardening-stress magnitude is identical in each grain and by choosing a value for k_1, the value for k_2 can be calculated analytically. While the hardening-stress magnitude is given by the experimental data, the k_1/k_2 combination has been fitted manually. The shape of the macroscopic stress-strain curve matches with the results of the reduced data RVE for the fitted k_1/k_2 combination.

It should be noted that fitting a two-dimensional EBSD scan to "three-dimensional" macroscopic experimental results is in fact a coarse approximation. The discrete volume-distribution of the macroscopic tension test is different from the area distribution in the EBSD scans. Nevertheless, the area distribution is used for the calculations in Section 3.5, because this model is phenomenological in its nature. Three-dimensional data is very rarely available or used in the literature.

The value found for k_1 is relatively low according to Delince et al. (2007) and Fang and Dahl (1995). However, the value for k_2 is well within the range of $7 - 22$ for ferritic steels given by Fang and Dahl (1995). The value for k_{HP} falls within the range of $15 - 18\,\mathrm{MPa}\sqrt{\mathrm{mm}}$ from the work of Resende and Bouvier (2013).

3.4 Analytical grain-size distributions

3.4.1 Numerical results and discussion

As depicted in Figure 3.6(a), (b), yielding starts in the largest grain that exhibits the smallest yield-strength. The Hall-Petch relation constitutes a grain-diameter dependency of the yield strength being $D^{-\frac{1}{2}}$. Therefore, the yield strength of larger grains varies significantly less with their grain size when compared to small grains. Larger grains yield rapidly one after another due to this behavior. With increasing external load, smaller grains start to yield until all grains yield. Then all grains either behave ideal plastic (Figure 3.6a) or harden with the same hardening modulus (Figure 3.6b).

The strain-partitioning parameter λ influences the stress and strain fluctuations and hence the macroscopic transition length. During the

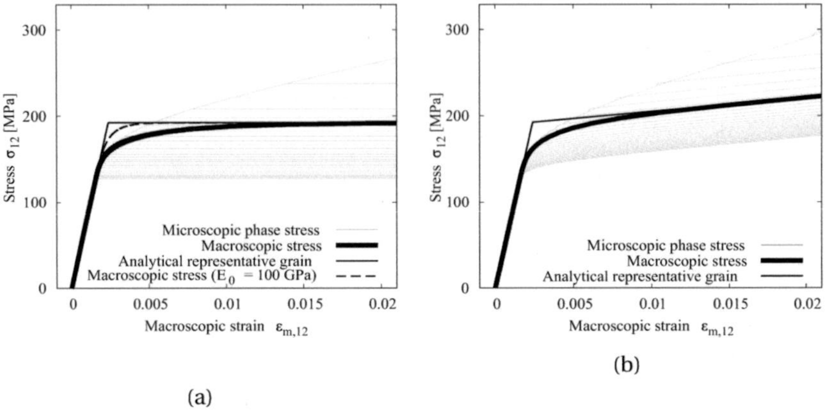

Figure 3.6: Stress-strain behavior of analytical log-normal distributions using different material models: (a) ideal plasticity; (b) linear isotropic hardening. Both for a parameter set of $\mu = 4\ \mu\text{m}$, $\sigma_\mu = 4\ \mu\text{m}$. Phases yield subsequently with different grain sizes within a regime. The λ-influence vanishes after the transition.

transition, stronger phases remain elastic and are only subject to a fraction of the macroscopic strain. The actual strain fraction is determined by λ and related to the ensemble average of the eigenstrains $\langle \varepsilon^{\mathrm{p}} \rangle$. When all phases are in a plastic state, the transition is complete and the macroscopic saturation stress $\bar{\boldsymbol{\sigma}} = \left\langle \boldsymbol{\sigma}^{\mathrm{f}} \right\rangle$ (Figure 3.6a) or, respectively, the macroscopic hardening modulus (Figure 3.6b) are not influenced by the parameter λ any more.

In the elastic regime, there are no stress or strain discrepancies for the special case of identical phase-stiffness tensors. Phases subjected to hardening behave figuratively more similar to the elastic case. Material hardening, therefore, creates an intermediate behavior, because it effectively reduces stress and strain fluctuations across the RVE and thus decreases the length of the transition regime (Figure 3.6b).

3.4.2 Estimated macroscopic saturation stress

In order to investigate the grain-size distribution effect, one can estimate the saturated constant yield stress (Figure 3.6a) as follows. For sufficiently high strains, it is a good assumption that all phases are in a plastic state. For ideal plasticity, the stress level in each phase is then determined a priori by the Hall-Petch relation. In this case, the macroscopic saturation-stress for a log-normal grain distribution $\bar{\sigma}^{\mathrm{f},\infty}$ can be exactly calculated by

$$\begin{aligned} \bar{\sigma}^{\mathrm{f},\infty} &= \int_0^\infty l\, c(D)\ \sigma^{\mathrm{f}}(D)\ \mathrm{d}D \\ &\overset{\text{e.g.}}{=} l\sigma_0 + l k_{\mathrm{HP}} \exp\left(\frac{m}{2}\left(S^2 m - 2M - 6S^2\right)\right). \end{aligned} \tag{3.6}$$

The factor l has its origin in the von Mises yield condition and depends on the load case ($l = 1$ for uniaxial tension, $l = \sqrt{3}$ for simple shear). This result explicitly confirms several outcomes of recent works, e.g., some of the main results of Berbenni et al. (2007) can be readily explained by use of Equation (3.6).

Figure 3.7 depicts the analytical estimation of the saturation stress for ideal-plastic phases with respect to the standard deviation σ_μ(Equation (3.6)). Two different mean grain sizes ($4\,\mu\mathrm{m}$, $10\,\mu\mathrm{m}$) are shown. At $\sigma_\mu = 0$, simple Hall-Petch stress of the mean grain size is obtained. The macroscopic saturation stress decreases with increasing dispersion. For a more realistic distribution of $\mu = 4\,\mu\mathrm{m}$ and $\sigma_\mu = 4\,\mu\mathrm{m}$ the saturation strength is notably decreased by $\sim\ 70\,\mathrm{MPa}$ (see Figure 3.1 for the distribution histogram).

The decreasing initial yield and saturation stress can be explained as an effect that is purely based on grain-size distribution. While the diameter distribution is shifted towards smaller grain-sizes with increasing dispersion, the volume distribution is shifted towards larger grains

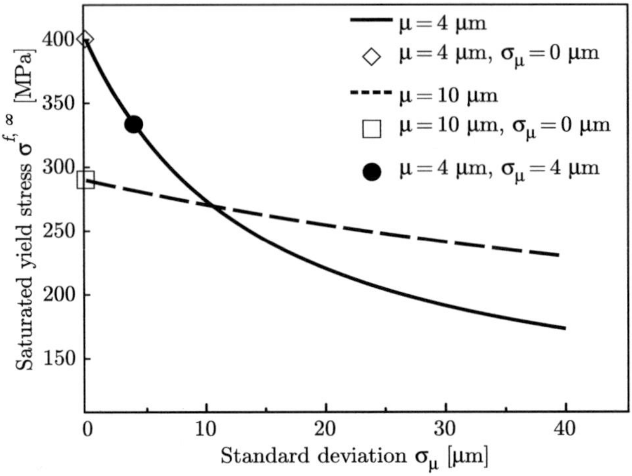

Figure 3.7: Analytical saturation stresses for an ideal plastic material model and a log-normal grain-size distribution

(Figure 3.1). Resembling rather the volume distribution than the diameter distribution, the macroscopic yield stress decreases with increasing dispersion; larger grains dominate the RVE response. The dispersion influence is larger for smaller average grain-sizes. This effect stems from a shape effect, inherent to the log-normal distribution.

It is remarkable that the yield-stress levels may even invert for higher standard deviations in Figure 3.7. Here, for standard deviations above 10 µm, the polycrystal with the smaller mean-diameter has a higher saturation stress. Rather than only the mean grain-size, the whole distribution is important for the saturation stress and hence for the complete macroscopical behavior. In more advanced material models, the single-grain behavior might be much more grain-size dependent, introducing a more involved relation between grain-size distribution and macroscopic behavior.

3.4.3 Representative grain size

Up to this point of the work, only the saturation stress of the RVE has been estimated. Raeisinia and Sinclair (2009) introduced the concept of a representative grain-size, exhibiting a stress-strain behavior similar to the complete RVE over the entire deformation. The volume-representative grain-size $\mu_{\mathrm{v}}^{\star}$ is calculated from the analytical saturation value of Equation (3.6),

$$\bar{\sigma}^{\mathrm{f},\infty} = \sigma_0 + \tfrac{k_{\mathrm{HP}}}{\sqrt{\mu^{\star}}}. \tag{3.7}$$

Assuming that the Hall-Petch relation holds on obtains

$$\mu_{\mathrm{v}}^{\star} = \mu \left(\frac{\sigma_{\mu}}{\mu^2} + 1 \right)^{\frac{9}{4}} \tag{3.8}$$

for the grain-size dependence. Equation (3.8) is only dependent on grain-size distribution parameters, i.e., the mean diameter μ and the standard deviation σ_{μ}. Since the experimental RVE makes use of a two-dimensional, EBSD-based RVE, a second “area-representative” grain-size must be calculated in analogy to Section 3.1,

$$\mu_{\mathrm{a}}^{\star} = \mu \left(\frac{\sigma_{\mu}}{\mu^2} + 1 \right)^{\frac{5}{4}} . \tag{3.9}$$

If one compares the stress-strain behavior of the representative grain with the behavior of the complete RVE (Figure 3.6), the agreement is quantitative before and after the transition (purely elastic to completely plastic, Section 3.4.1). The difference between the results for the representative grain and the results for the RVE is more pronounced for higher strain fluctuations. For equal phase strains ($\lambda \to 0$), the transition is comparably short and the best agreement is obtained. For in-

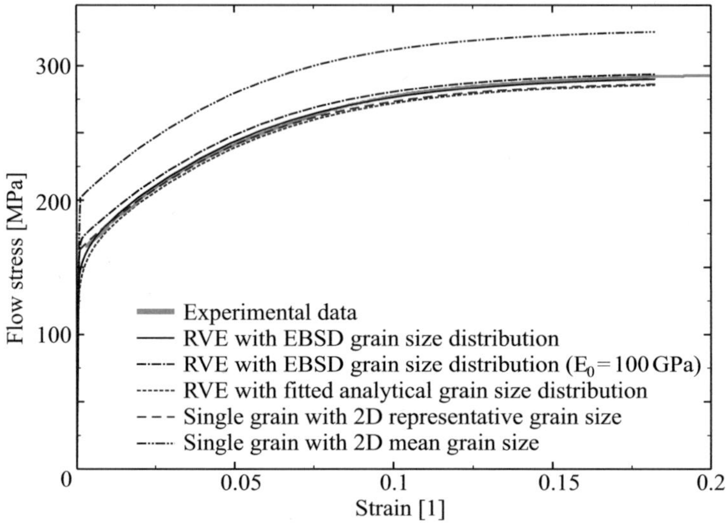

Figure 3.8: Stress-strain diagram of the different RVEs using the material model from Kocks and Mecking (2003) compared to the experimental data. The strain-partitioning is close to the iso-strain or Taylor case.

termediate to equal-stress states ($\lambda \rightarrow 1$), the transition becomes more pronounced and the transition length increases.

The microscopic and macroscopic stress-strain curves in Figure 3.6 (a) and (b) have different geometric shapes (a distinct kink or, respectively, a smooth transition). This is mainly due to simplifications of the first two material models. A more realistic material model with a smooth transition from elastic to plastic behavior leads to a better accordance between the representative grain results and the RVE results (Figure 3.8). For this case, microscopic and macroscopic stress-strain curves are geometrically similar. The agreement is almost quantitative for the Kocks and Mecking (2003) model (Section 3.5).

3.5 Experimental grain-size distributions – results and discussion

Five different RVE simulations have been carried out in order to investigate the effects in question (Figure 3.8):

1. two experimental RVEs based on the reduced EBSD data set, differing only in the strain-partitioning parameter λ;
2. an analytical log-normal RVE with the grain-size distribution parameters $\mu = 18\ \mu\text{m}$ and $\sigma_{\mu} = 312\ \mu\text{m}$, taken from the EBSD scan;
3. and finally two RVEs composed of equally-sized grains with either the mean diameter ($D = \mu$) or equally-sized grains with the representative diameter ($D = \mu_{\text{a}}^{\star} = 41\ \mu\text{m}$).

In view of the simulations that differ only in λ (EBSD-RVE), it is evident that this parameter only affects the first percent of the stress-strain curve. The material exhibits a distinct yield point. For higher values of λ, the transition length increases which, macroscopically, causes continuous yielding. If the strains are close to the iso-strain assumption, the prediction from the representative grain-size becomes most accurate. The microscopic stress evolution does not reflect stress and strain inhomogeneities from, e.g., texture development Jöchen (2013).

There is only a minor difference between the two stress-strain curves from the EBSD-based simulation results and the analytical RVE with a fitted analytical log-normal distribution. This small difference may be explained by the probability-density deviance from the analytical log-normal distribution in the region up to $20\ \mu\text{m}$ (Figure 3.3). The shift of the probability-density to lower grain sizes acts in conjunction with the nonlinear Hall-Petch behavior (scaling by $D^{-\frac{1}{2}}$). This results in the slightly increased macroscopic stress state compared to the fitted log-normal grain-size distribution.

The stress-strain response of the RVE with the representative diameter $\mu_a^\star$, matches the experimental RVE behavior well. The formula matches the values of analytical RVE calculations, too. This is an indicator for a sufficient discretization quality within the RVE. Raeisinia and Sinclair (2009) explained the robustness of the method based on Equation (3.9). They found that a spectrum of grains sizes close to the analytical representative diameter is in good agreement with the macroscopic behavior.

For identical material parameters, the macroscopic behavior of a homogeneous polycrystal with the average diameter μ is significantly differing from the heterogeneously distributed polycrystal. This highlights the applicability of the representative grain-size concept for polycrystalline metals.

3.6 Conclusions on the influence of the grain-size distribution

Albeit a significant influence of the grain-size distribution on the macroscopic material behavior has been observed in this section, the average diameter is used in Chapter 6 with the following arguments: Firstly, for the work-hardening of DP steels, much more micromechanical information or precision is lost from actual model uncertainties than from an incorrect representative grain-size. Secondly, the grain-size dispersion effect is easily hidden by parameter fitting. If one chooses the mean diameter instead of the more physically representative diameter, all remaining parameters will adapt. This effectively covers the grain-size distribution effect. Thirdly, for the proposed model, the initial yield-stress is not determined from a grain size relation but fitted to experimental data. And finally, an extension of the Mughrabi et al. (1986) composite model to more than two phases would complicate the proposed model disproportionately.

Chapter 4

Grain-level plasticity in dual-phase steels – full-field RVE simulations

In this section, the material behavior of DP steels is studied at the grain-level by simulating EBSD-based microstructures with a crystal-plasticity constitutive model[1]. The finite-element full-field simulations facilitate local and statistical investigations. At the time when the simulations where carried out, no experimental data was available for the DP600 steel (Salzgitter Flachstahl GmbH, Germany) that is mainly investigated in Chapter 7. Therefore, the simulations in this section are based on available data from a different, DP600v, steel manufactured by voestalpine AG, Austria.

4.1 Finite element modeling based on EBSD data

Martensite identification procedure. To examine the behavior of DP steels, a finite-element RVE is derived from real EBSD data of a voestalpine® DP600v steel. In the first step, constituents are identified by a newly proposed procedure. The procedure is similar to the

[1] This section is reproduced from Rieger and Böhlke (2015).

C_{DP}	P	Mn	S	Cr	Cu	Mo	Si	Ni	C_α	C_γ
0.094	0.01	1.5	0.0054	0.77	0.061	0.003	0.1	0.024	0.004	0.45
Wasilkowska et al. (2006); Larour et al. (2013)									Ramazani et al. (2013)	

Table 4.1: Chemical composition of voestalpine® DP600v dual-phase steel in wt.% (remainder is Fe); $C_\gamma = (C_{DP} - (1 - c_M)\, C_\alpha)\,/c_M$ from the rule of mixtures.

one of Choi et al. (2013) but based on standard image manipulation techniques in the work at hand.

Using the rule of mixtures, the carbon content is estimated to be about $C_M \simeq 0.45$. Thus the investigated DP600v steel will mainly exhibit lath martensite (Krauss, 1999). High densities of tangled dislocations from transformation and accommodation effects are characteristics of lath martensite. Consequently, the procedure identifies high-distortion areas in the EBSD data as initial martensite, i.e., measurement errors or single measurement grains (see also Wilson et al., 2001). The identified martensite-areas are then contracted (eroded) by image-manipulation algorithms in order to eliminate random noise and grain boundaries. To restore the initial martensite grain-geometry, a „blow-up“ (dilatation) step is applied to the remaining martensite and a constant artificial crystallographic orientation is assigned to the martensite identified in this way. After this processing, the modified EBSD data set is evaluated with MTEX (Bachmann et al., 2011). The crystallographic orientation of martensite inclusions is ignored for the simulations because the EBSD measurements were prone to errors (which in turn enabled the proposed identification procedure).

Simulation setup. To carry out the finite-element simulations, the RVE geometry (Figure 4.1) is imported into ABAQUS. The crystal-plasticity material model (Section 4.2.1) is assigned to each ferrite grain with an individual grain size and orientation. Martensite regions are identically modeled by the model that is detailed in Section 4.2.3. A

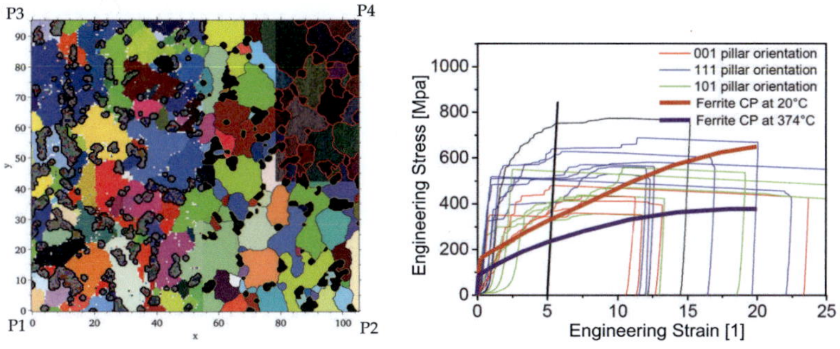

(a) (b)

Figure 4.1: Comparison of the original EBSD data with the martensite grains identified by the proposed algorithm (a). The overlay on the upper right side shows the finite-element mesh used in the simulations. Micro-pillar compression tests and ferrite constituent behavior within the range of the micro-pillar compression test (Wenk and Schreijäg, 2013).

total of 195 ferrite grains and 161 martensite grains ($20\,\%$ martensite content, $c_{\mathrm{M}} = 0.2$) is simulated in the RVE.

The two-dimensional RVE data is extruded by a small amount compared to the RVE dimensions and discretized by $268 \cdot 10^3$ fully integrated ABAQUS C3D8 elements (simulations took around six to ten hours on 12 CPUs). A comparison calculation for a finer mesh shows only negligible changes in the resulting stress field and the macroscopic behavior. It is concluded that the mesh is sufficiently fine for the purpose of the work at hand.

Periodic and kinematic boundaries according to Danielsson et al. (2007) are prescribed in x- and y-direction; this allows for strain fluctuations and avoids localizations at the RVE boundary. In the work of Danielsson et al. (2007), the macroscopic displacement gradient $\bar{\boldsymbol{H}}$ is used to express the connection between the displacement at two arbitrary periodic points $\boldsymbol{u}^{+}$ and $\boldsymbol{u}^{-}$

$$\boldsymbol{u}^{+} - \boldsymbol{u}^{-} = \bar{\boldsymbol{H}} \left[\hat{\boldsymbol{X}}^{+} - \hat{\boldsymbol{X}}^{-} \right]. \tag{4.1}$$

From this formula the equation boundary conditions are implemented in ABAQUS with regard to four reference points $\boldsymbol{u}^{P1}, \boldsymbol{u}^{P2}, \boldsymbol{u}^{P3}, \boldsymbol{u}^{P4}$ (Figure 4.1) by enforcing

$$\begin{aligned} \boldsymbol{u}_{x^+} - \boldsymbol{u}_{x^-} &= \boldsymbol{u}^{P2} - \boldsymbol{u}^{P1}, \\ \boldsymbol{u}_{y^+} - \boldsymbol{u}_{y^-} &= \boldsymbol{u}^{P4} - \boldsymbol{u}^{P1}, \\ \boldsymbol{u}^{P3} - \boldsymbol{u}^{P4} &= \boldsymbol{u}^{P2} - \boldsymbol{u}^{P1}. \end{aligned} \tag{4.2}$$

Mesh nodes with identical $x-$ and $y-$coordinates are constrained to identical displacements on the RVE boundary. The macroscopic deformation was defined by applying displacement boundary conditions for the four reference points: $u_x^{P1} = u_y^{P1} = 0$, $u_y^{P2} = 0$ and $u_x^{P4} = 0$ (which leaves $u_{x,y}^{P3}$ unrestricted) for the quenching step. In addition, the displacement $u_x^{P2} = \bar{u}_{13\%}$ is prescribed during the uniaxial tension step. In z-direction, the displacement boundary condition is $u_z = 0$ for all nodes on the rear plane. All simulation results lie in between the stiff plain-strain and soft plain-stress extremal cases. The complete boundary-condition definition may be used for both the volumetric expansion and the uniaxial-loading step.

Formation of GNDs is simulated by a virtual quenching step for which the homogeneous RVE-temperature is linearly lowered from the martensite-start temperature $M_\mathrm{s} = 374\,^\circ\mathrm{C}$ to room temperature at $20\,^\circ\mathrm{C}$. In the second load step, macroscopic uniaxial-tension or uniaxial-compression is linearly prescribed in horizontal direction. The RVE-results for loading along the vertical direction exhibit a very similar macroscopic behavior.

4.2 Constitutive models for individual constituents

4.2.1 Ferrite material model at room temperature

For the ferritic matrix, the geometrically nonlinear constitutive model is based on an elasto-viscoplastic crystal-plasticity formulation for single crystals (Böhlke, 2001; Jöchen and Böhlke, 2012). The reader is referred to these works for more details; only a summary of the model is given, here, for brevity.

The model assumes small elastic-strains, and finite plastic-strains and rotations. With $\boldsymbol{F} = \boldsymbol{F}_{\mathrm{e}}\boldsymbol{F}_{\mathrm{p}}$, the deformation gradient is decomposed into an elastic and plastic part. The elastic part $\boldsymbol{F}_{\mathrm{e}}$ accounts for lattice distortion, while the plastic part $\boldsymbol{F}_{\mathrm{p}}$ accounts for crystallographic slip. Furthermore, the elastic law for the Kirchhoff stress tensor $\boldsymbol{\tau} = \det(\boldsymbol{F})\,\boldsymbol{\sigma}$ is assumed to be

$$\boldsymbol{\tau} = \boldsymbol{F}_{\mathrm{e}}\tilde{\mathbb{C}}\left[\boldsymbol{E}_{\mathrm{e}}\right]\boldsymbol{F}_{\mathrm{e}}^{\mathsf{T}} \tag{4.3}$$

with Green´s strain tensor $\boldsymbol{E}_{\mathrm{e}} = \left(\boldsymbol{F}_{\mathrm{e}}^{\mathsf{T}}\boldsymbol{F}_{\mathrm{e}} - \boldsymbol{I}\right)$. Using the orthonormal base-tensors $\boldsymbol{B}_{\alpha}$, the cubic material-stiffness tensor is given by

$$\tilde{\mathbb{C}} = \begin{bmatrix} C_{1111} & C_{1122} & C_{1122} & 0 & 0 & 0 \\ & C_{1111} & C_{1122} & 0 & 0 & 0 \\ & & C_{1111} & 0 & 0 & 0 \\ & & & 2C_{1212} & 0 & 0 \\ & \mathrm{sym.} & & & 2C_{1212} & 0 \\ & & & & & 2C_{1212} \end{bmatrix} \boldsymbol{B}_{\alpha} \otimes \boldsymbol{B}_{\beta}. \tag{4.4}$$

Due to being symmetric itself and of cubic material symmetry, the material-stiffness tensor is described by three elastic constants (Jöchen and Böhlke, 2012).

Based on the Schmid tensor in the undistorted configuration $\tilde{\boldsymbol{M}}_\alpha$, the time evolution of $\boldsymbol{F}_\mathrm{p}$ is given by the rate-dependent flow rule,

$$\dot{\boldsymbol{F}}_\mathrm{p}\boldsymbol{F}_\mathrm{p}^{-1} = \sum_\alpha \dot{\gamma}_\alpha \tilde{\boldsymbol{M}}_\alpha. \tag{4.5}$$

Slip rates are given by means of the reference shear-rate $\dot{\gamma}_0$ and the critical resolved shear stress τ^C, which, in turn, is defined as

$$\dot{\gamma}_\alpha = \dot{\gamma}_0 \mathrm{sgn}\left(\tau_\alpha\right)\left|\frac{\tau_\alpha}{\tau^\mathrm{c}}\right|^m. \tag{4.6}$$

Here, m quantifies the strain-rate sensitivity, chosen to be very small. Only 24 bcc slip-systems from the combinations of the $\{110\}\left\langle 111\right\rangle$ and $\{112\}\left\langle 111\right\rangle$ slip-system families are considered. The resolved shear-stress is given by

$$\tau_\alpha = \mathrm{dev}(\boldsymbol{T}_\mathrm{e}) \cdot \tilde{\boldsymbol{M}}_\alpha \tag{4.7}$$

and is computed using the Mandel stress tensor

$$\boldsymbol{T}_\mathrm{e} = \det\left(\boldsymbol{F}_\mathrm{e}\right)\boldsymbol{F}_\mathrm{e}^\mathsf{T}\boldsymbol{\tau}\boldsymbol{F}_\mathrm{e}^{-\mathsf{T}}. \tag{4.8}$$

The rate of accumulated plastic-slip is computed by integrating

$$\dot{\gamma} = \sum_\alpha \left|\dot{\gamma}_\alpha\left(\tau_\alpha, \tau^C\right)\right| \tag{4.9}$$

over time with the initial condition $\gamma\left(t = 0\right) = 0$.

In crystal-plasticity theories, material hardening is commonly described with, e.g., a Voce model (Kocks and Mecking, 2003)

$$\dot{\tau}^\mathrm{C}\left(\tau_\alpha, \tau^C\right) = \Theta_0\left(1 - \frac{\tau^C}{\tau^C_\mathrm{V}}\right)\dot{\gamma} \tag{4.10}$$

with the initial hardening-modulus Θ_0. There exists a direct relationship between the phenomenological Voce law and the Kocks-Mecking formulation in Equation (4.23). Ferrite yield curves are, additionally, dependent on the current temperature and the grain size. At lower temperatures, interstitial-free ferrite steels exhibit lower yield strengths and less-pronounced work hardening. Nevertheless, the geometric (exponential) shape of the stress-strain curve remains approximately the same (Ramazani et al., 2013). The reasoning for the coming modifications, is based on the idea that work hardening in ferrite is dominated by dislocation interactions.

Temperature dependence. A simple extension was applied to the initial shear-stress τ_0^{C} and the saturated critical shear-stress $\tau_{\mathrm{V}}^{\mathrm{C}}$. Therefore, a Hall-Petch term is added to the original critical shear stress to model the grain-size influence on the initial yield-stress. Secondly, a linear temperature-scaling is introduced through $f^{\mathrm{F}}(\theta)$. The empirical function $f^{\mathrm{F}}(\theta)$ is described at the end of Section 4.2.2. With all these changes, the initial value of the critical resolved shear stress is then given by the expression

$$\tau_0^{\mathrm{C}} := \tau^{\mathrm{C}}(0) = \left(\tau_{00}^{\mathrm{C}} + \frac{k_{\mathrm{HP}}^{\tau}}{\sqrt{D}}\right) f^{\mathrm{F}}(\theta)\,. \tag{4.11}$$

The critical Voce-stress is similarly extended to model the increased dislocation-recovery at higher temperatures, i.e.,

$$\tau_{\mathrm{V}}^{C}\left(\tau_\alpha, \tau^{C}\right) = \left(\tau_{\mathrm{V}0}^{C} + \frac{k_{\mathrm{HP}}^{\tau}}{\sqrt{D}}\right) f^{\mathrm{F}}(\theta)\,. \tag{4.12}$$

With this model, work-hardening (the difference between initial yield and higher yield stresses for increasing plastic deformation) is independent of the grain size. This is observed, e.g., in Gao et al. (2014) for grain sizes as small as $\sim 2\,\mu\mathrm{m}$ in interstitial-free steel. The cur-

Parameter	Symbol	Value	Reference
Ferrite stiffness	C_{1111}	231.5 GPa	Kim and Johnson (2007)
	C_{1122}	135 GPa	
	C_{1212}	116 GPa	
Strain rate sensitivity	m	20	Jöchen and Böhlke (2012) for DC04 steel
Reference rate	$\dot{\gamma}_0$	$1 \cdot 10^{-3}\,\mathrm{s}^{-1}$	
Initial critical shear	τ_0^{C}	45 MPa	fitted to macroscopic DP steel behavior
Asymptotic crit. shear	$\tau_{\mathrm{V0}}^{\mathrm{C}}$	230 MPa	
Initial hardening modul	Θ_0	1500 MPa	
Hall-Petch factor	k_{HP}^{τ}	$50\,\mathrm{MPa}\sqrt{\mu\mathrm{m}}$	Similar to Delince et al. (2007)

Table 4.2: Material parameters for the ferrite crystal plasticity model

rent critical shear-stress, prior to any hardening, is calculated at the integration-point level by $\tau^{\mathrm{C}} = \tau_0^{\mathrm{C}} + \Delta\tau_n$. The term $\Delta\tau_n$ denotes the total stress-increment from work-hardening that was present at the previous time increment n. It is, thus, a measure for dislocations that were accumulated in the material so far. Using this formulation, the accumulated dislocation-density (or $\Delta\tau_n$) remains constant for arbitrary temperature changes. In general, temperature changes are not negligible for the dislocation-density evolution. In spite of this, the presented approach is still applicable for this particular investigation as only quenching is considered.

To identify the initial critical shear-strength at room temperature, the empirical formula of Rodriguez and Gutiérrez (2003); Resende and Bouvier (2013) is scaled by the Taylor factor M. Then the critical shear-strength is given in MPa by

$$\tau_0^{\mathrm{C}}\,[\mathrm{MPa}] = \frac{1}{M}(5000\cdot\mathrm{C}_\alpha + 42\cdot\sqrt{\mathrm{Mn}} + 750\cdot\mathrm{P} + 60\cdot(\mathrm{Si}+\mathrm{Cr}) + 80\cdot\mathrm{Cu} + 11\cdot\mathrm{Mo}). \tag{4.13}$$

The parameters $\tau_{\mathrm{V0}}^{\mathrm{C}}$ and Θ_0 are fitted to match the range of the mechanical response given by micro-pillar compression tests (see Figure 4.1b).

4.2.2 Austenite/martensite material models at room temperature

For the combined austenite-martensite material model the standard von Mises material-behavior is applied (Equation 1.8).

Martensite model. The martensite hardening-model is based on the works of Rodriguez and Gutiérrez (2003), i.e., the flow stress evolution is given by the expression

$$\sigma_{\mathrm{y}}^{\mathrm{M}} = \sigma_0 + MG\alpha\sqrt{b}\sqrt{\frac{1-\exp\left(-Mk_{\mathrm{r}}\varepsilon^{\mathrm{p}}\right)}{k_{\mathrm{r}}L}}. \tag{4.14}$$

The corresponding constant, the initial yield-strength σ_0, is estimated (in MPa) according to Thomser et al. (2009),

$$\sigma_0\,[\mathrm{MPa}] = (3065\cdot \mathrm{C}_\gamma - 161) + 42\cdot\sqrt{\mathrm{Mn}} + 750\cdot\mathrm{P} + 60\cdot\mathrm{Si} + 80\cdot\mathrm{Cu} + 11\cdot\mathrm{Mo}. \tag{4.15}$$

The chemical alloy-composition is given in weight percent. Here, dislocation interaction is related to the constant α. Additional material parameters are present in the form of the Taylor factor M, the shear modulus G, and the magnitude of the Burger´s vector b. The recovery rate k_{r}, the mean-free dislocation path L, and the equivalent plastic-strain ε^{p} govern the work-hardening rate.

Austenite model. The austenite flow curve is given by Ludwik´s equation $\sigma_{\mathrm{f}}^{\mathrm{A}} = a_1 + (a_2\varepsilon^{\mathrm{p}})^{0.4}$. A more detailed description of the modeling process and the parameters for austenite and martensite in Table 4.3 can be found in Ramazani et al. (2013). Note that only isotropic material-models were used in their work.

4.2.3 Austenite-martensite phase transformation modeling

The already expanded DP microstructure from the EBSD scan is taken as a starting point for the phase transformation. This is motivated by

Parameter	Symbol	Value	Reference
Disl. interaction	α	0.3	Ramazani et al. (2013)
Taylor factor	M	3	
Shear modulus	G	$80\,\mathrm{GPa}$	
Magnitude of Burgers vector	b	$2.5 \cdot 10^{-10}\,\mathrm{m}$	
Recovery rate	k_r	41	
Disl. mean free path	L	$3.8 \cdot 10^{-8}\,\mathrm{m}$	
Austenite initial yield	a_1	$650\,\mathrm{MPa}$	
Austenite hardening	a_2	$1200\,\mathrm{MPa}$	
Martensite start	M_s	$374\,^\circ\mathrm{C}$	
Martensite finish	M_f	$234\,^\circ\mathrm{C}$	
Transformation	k	$3.29 \cdot 10^{-2} K^{-1}$	fitted to $M_\mathrm{s}/M_\mathrm{f}$
Ferrite vol. extension	α_θ^F	$1.2 \cdot 10^{-5} K^{-1}$	Ramazani et al. (2013)

Table 4.3: Material parameters for the isotropic austenite-martensite model

the overall small geometric change during the phase transformation ($\sim 3.4\,\%$). At the martensite start temperature M_s, non-ferritic areas are assumed to be purely austenitic. These areas transform completely to martensite until the martensite finish-temperature M_f is reached.

During the austenite-martensite phase transformation, a mixture of two phases is present. The complex interactions during the phase transformation are simplified to follow the rule of mixtures for austenite and martensite (Liedl et al., 2002; Ramazani et al., 2013), i.e.,

$$\sigma_\mathrm{f}^{\mathrm{A/M}} = \xi \cdot \sigma_{\mathrm{f},M_\mathrm{f}}^{\mathrm{M}} + (1-\xi) \cdot \sigma_{\mathrm{f},M_\mathrm{s}}^{\mathrm{A}}. \tag{4.16}$$

Equation (4.16) interpolates the mixture´s flow curve between purely austenitic behavior at $M_\mathrm{s} = 374\,^\circ\mathrm{C}$ and purely martensitic behavior at $M_\mathrm{f} = 234\,^\circ\mathrm{C}$, assuming equal strains in both phases. The martensite fraction during transformation is calculated by the standard Koistinen and Marburger (1959) equation that was fitted to the given temperature interval, i.e.,

$$\xi = 1 - \exp\left(-k\left(M_\mathrm{s} - \theta\right)\right). \tag{4.17}$$

During the transformation, the surrounding ferrite-matrix poses compatibility constraints on the developing martensite. This leads to the formation of several different martensite-variants within a single martensite island. Such a behavior is characteristic for lath martensite (Krauss, 1999). It is commonly believed, that the total of the shear components from the different martensite-variants, approximately, vanishes (Kadkhodapour et al., 2011). In Ramazani et al. (2013), the resulting transformational volume-expansion is modeled by an artificial thermal expansion coefficient. Analogously, a linear martensite-expansion of $3.4\,\%$ is modeled during the phase transformation in the work at hand. A volume expansion of $3.4\,\%$ is higher than the specific volume difference between austenite and martensite. It is, however, in line with experimental results (Kadkhodapour et al., 2011). Since the chosen volume strain increment of $\delta = 0.034$ is equivalent to a spherical transformation strain of $\boldsymbol{\varepsilon}^{\theta} = \delta/3\boldsymbol{I}$, the artificial volume-expansion coefficient during the phase transformation can be calculated from the experimental coefficient for ferrite by

$$\bar{\alpha}_{\theta}^{\mathrm{A/M}} = \alpha_{\theta}^{\mathrm{F}} - \frac{\delta}{3}\frac{1}{\theta} = -6.895 \cdot 10^{-5}\frac{1}{\mathrm{K}}. \tag{4.18}$$

At the end of the martensite transformation, the volume expansion coefficients of ferrite and martensite are equal in the simulation. In order to circumvent convergence problems, a smooth transition in the range of around $M_{\mathrm{f}} \pm 5\,\mathrm{K}$ is implemented as ABAQUS "user expansion subroutine" (UEXPAN). The transition is based on the tangent hyperbolicus function. This approach gives an artificial volume-expansion coefficient by

$$\alpha_{\theta}^{\mathrm{A/M}} = \frac{1}{2}\left(\bar{\alpha}_{\theta}^{\mathrm{A/M}} + \alpha_{\theta}^{\mathrm{F}} + \left(\bar{\alpha}_{\theta}^{\mathrm{A/M}} - \alpha_{\theta}^{\mathrm{F}}\right)\tanh\left(\frac{1}{2}\left(\theta - M_{\mathrm{f}}\right)\right)\right). \tag{4.19}$$

In order to extrapolate the material behavior from room temperature ($\sigma_{\mathrm{f},RT}$) to elevated temperatures, the complete flow-curves are scaled according to the procedure in Ramazani et al. (2013). The underlying data was taken from comparable austenitic, ferritic and martensite steels (Ramazani et al., 2013). The resulting scaling equations are linear in the temperature (θ given in °C), i.e., for ferrite it is

$$f^{\mathrm{F}}(\theta) = (-0.0013 \cdot \theta + 1.026)\,, \tag{4.20}$$

for martensite the relation is

$$\sigma_{\mathrm{f}}^{\mathrm{M}}(\theta) = (-0.000327 \cdot \theta + 1.007)\,\sigma_{\mathrm{f},RT}^{\mathrm{M}} \tag{4.21}$$

and austenite is, analogously, given by

$$\sigma_{\mathrm{f}}^{\mathrm{A}}(\theta) = (-0.001004 \cdot \theta + 1.019)\,\sigma_{\mathrm{f},RT}^{\mathrm{A}}. \tag{4.22}$$

Even if no plastic deformation was present in the material ($\Delta\tau_n = 0\,\mathrm{MPa}$, see Equation (4.11)) the critical shear-strength $\tau_{\mathrm{V}}^{\mathrm{C}}$ is still observed to increase in experiments. In order to represent this behavior, the initial shear strength τ_0^{C} is scaled according to Equation (4.11). Consequently, the accumulated dislocation-density remains constant for arbitrary temperature changes. It should be noted that such a behavior is only applicable for rapidly- and monotonically-decreasing temperatures (as during the simulated quenching process). For the work at hand, the time-dependent creep effects are negligible and, for this special case, the dislocation density increases monotonically.

The work-hardening rate and the saturated critical shear-stress both decrease at elevated temperatures. This results from the more pronounced dislocation movement due to the increasing lattice oscillation at these temperatures. In consequence, the recovery rate increases as well. To represent this behavior in the model, the saturation

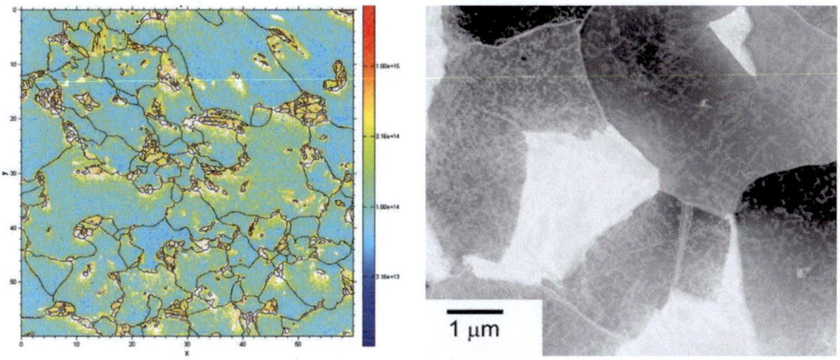

(a) (b)

Figure 4.2: GND density estimation from EBSD misorientation for the investigated DP600v (Wenk and Schreijäg, 2013; a); scanning transmission electron microscopy image of a DP600 steel with $c_{\mathrm{M}} = 0.3$ (Asadi et al., 2012; b). In Figure (a), the dislocation density gradually increases from low values in the grain interior ($1 \cdot 10^{14}\,\mathrm{m}^{-2}$) towards ferrite-ferrite grain-boundaries ($5 \cdot 10^{14}\,\mathrm{m}^{-2}$), but even more so towards ferrite-martensite phase-boundaries ($1 \cdot 10^{15}\,\mathrm{m}^{-2}$). After the phase transformation, areas of higher dislocation density are present.

stress $\tau_{\mathrm{V}}^{\mathrm{C}}$from Equation (4.12) is scaled by $f^{\mathrm{F}}(\theta)$ as well. With regard to the Taylor relation in Equation (4.23), scaling $\tau_{\mathrm{V}}^{\mathrm{C}}$ down like this is equivalent to an increased dislocation storage at lower temperatures. A lower percentage of generated dislocations is recovered dynamically.

4.3 Microscopic initial material state

It is widely accepted that for certain temperature ranges, the macroscopic work-hardening evolution can be described by (e.g. Rauch et al., 2011)

$$\sigma = M \left(\tau_0 + \alpha G b \sqrt{\rho}\right). \tag{4.23}$$

There is a direct relationship between the description based on the phenomenological Voce-law and the formulation in Equation (4.23) (Mecking, 2001). The average dislocation-density can, therefore, be estimated for the isotropic hardening which in turn is applied in the crystal-plasticity model, i.e.,

$$\rho = \left(\frac{\tau^{C} - \tau_0^{C}}{\alpha G b}\right)^2 . \tag{4.24}$$

Experimental results. Figure 4.2 shows the GND-density estimation from EBSD misorientation scans as performed by Schreijäg (2013) for the considered DP600v steel. The dislocation density is generally lower in the interior of ferrite grains ($1 \cdot 10^{14}\,\mathrm{m}^{-2}$) because the ferrite grain interior recovers during the heat treatment. Similar to the results of Korzekwa et al. (1984); Calcagnotto et al. (2010a); Dillien et al. (2010); Kapp et al. (2011), the dislocation density gradually increases towards ferrite-ferrite interfaces ($5 \cdot 10^{14}\,\mathrm{m}^{-2}$). It increases even more so towards ferrite-martensite interfaces ($1 \cdot 10^{15}\,\mathrm{m}^{-2}$). The same trend is observed for the EBSD-estimated dislocation-density during the subsequent reloading experiment of the DP600 steel (the steel characterized in Chapter 5 that is simulated for Chapter 7). The major difference between the two DP steels lies in the martensite arrangement: martensite is dispersed and exhibits small grain-sizes in the DP600 steel, however, the DP600v steel exhibits block structured martensite islands. Therefore, the aforementioned initial gradient is difficult to observe.

The amount of plastic deformation in the GND layer is influenced by the spatial arrangement of martensite islands relative to each other. Narrowly-positioned martensite islands induce higher deformations in the ferrite matrix because the deformations from different martensite islands interact. Therefore, the spatial arrangement of martensite islands, e.g., smaller overall grain-sizes or network-like martensite

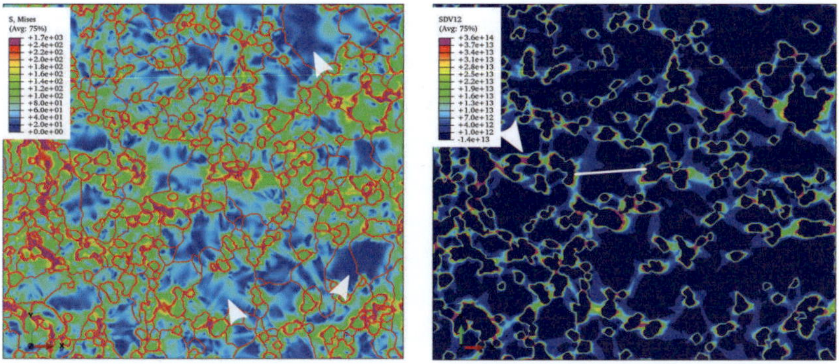

(a) (b)

Figure 4.3: Equivalent von Mises stress in the RVE after quenching (a); estimated dislocation density in the full-field RVE after quenching (b). The path for the line-scan in Section 4.5.2 is indicated by the white segment in Figure (b).

structures, imposes an important microstructural characteristic (Balliger and Gladman, 1981). Calcagnotto et al. (2010a) argue, that small martensite-inclusions also lead to layers of high GND density. Their carbon content tends to be higher which, in turn, leads to a larger volume expansion. In principle, this means that the initial dislocation-density in DP steels with high-carbon martensite should be higher than in DP steels with low-carbon martensite.

Figure 4.3(a) shows the finite-element results of the full-field RVE after the austenite-martensite phase transformation has been simulated, i.e, at $0\,\%$ macroscopic strain. DP steel is subjected to eigenstresses and plastic deformations after the quenching step. Transformational volume-expansion has caused incompatibility effects, i.e., one observes an increased dislocation-density in the ferrite matrix within a layer around each martensite island. The ferrite grain-interior is already subjected to elastic von Mises stresses in the range of $60{-}100\,\mathrm{MPa}$ before any macroscopic load is applied.

Graded influence zone. The material state in the RVE inside the ferrite GND-layer shows a graded structure as will be thoroughly discussed in Section 4.5.1 with regard to Figure 4.8. At ferrite-martensite interfaces, the highest deformation is observed. There, both the von Mises stress and the dislocation density ($\sim 10^{13}\,\mathrm{m}^{-2}$)are maximal. With increasing distance from ferrite-martensite interfaces, the GND density decreases down to values where almost no effect of the phase transformation is recognizable. Microhardness measurements of Kadkhodapour et al. (2011) and Tsipouridis et al. (2011) confirm this general trend. They also find the gradient to be of exponential and decay-like shape around martensite islands.

In this work, a dislocation density of $1 \cdot 10^{12}\,\mathrm{m}^{-2}$is defined as threshold for the influence of the GND layer zone. This value is equivalent to a work hardening of about $\Delta\tau = 7\,\mathrm{MPa}$. As depicted in Figure 4.3(b), the GND layer in this work exhibits widths from $w_{\mathrm{GND}} = 0.5\,\mu\mathrm{m}$ to $1.5\,\mu\mathrm{m}$. These values are well within the range of experimental results from recent publications, i.e., see the results from the literature in Table 4.4.

The initial dislocation-density in the RVE after the phase transformation is relatively low compared to the experimental data in Figure 4.2 and compared to values from literature (Table 4.4). A saturated critical shear-stress of $\Delta\tau^{\mathrm{C}} = 85\,\mathrm{MPa}$ for the ferrite model is equivalent to a dislocation density of $9 \cdot 10^{14}\,\mathrm{m}^{-2}$. Thus, in principle, such high dislocation-density values can be modeled with the employed ferrite model. To investigate this effect, simulations have been carried out with an increased transformational volume-expansion of $5\,\%$. This is about the highest experimentally observed value of volume expansion. However, the results do not show a dislocation-density increase by two orders of magnitude.

Martensite grain-coverage. Increasing the amount of ferrite grain boundary that is covered by martensite is commonly found to raise the

w_{GND} [µm]	$\rho_{\mathrm{FI}} - \rho_{\mathrm{FM}}$ $[10^{13}\,\mathrm{m}^{-2}]$	**Source**	**Material**	D [µm]	V_{M} [%]	**Reference**
< 1	$60 - 300$	TEM	DP400/900	~ 7	10	Gardey et al. (2005)
~ 1	$2.5 - 25$	EBSD	DP500/900	1.4	23	Calcagnotto et al. (2010a)
~ 2	6	EBSD	DP400/600	3.5	15	Dillien et al. (2010)
6	–	Hardness	DP	~ 30	45	Tsipouridis et al. (2011)
1.5	–	Hardness	DP400/700	2	–	Kadkhodapour et al. (2011)
0.7	$2 - 13$	EBSD	DP400/900	4.5	35	Ramazani et al. (2013)
$0.5 - 2$	$7 - 50$	EBSD	DP300/600	6.3	18	this work
~ 1	$0.1^{(*)} - 7$	Simulation				

Table 4.4: Comparison of ferrite-martensite phase boundary width and dislocation densities. For $^{(*)}$ $\rho_{\mathrm{FI}} = 1 \cdot 10^{12}\,\mathrm{m}^{-2}$ is manually defined as threshold value for the influence of the GND layer zone. Transmission-electron microscopy measures statistically-stored dislocations. Thus, the observed dislocation densities are higher than the values estimated from EBSD scans. Dislocation densities from the simulations are smaller, but the width of the influence zone w_{GND} is in line with experimental data.

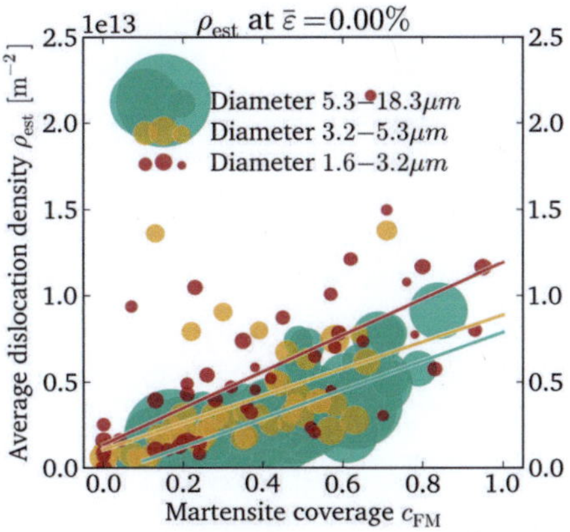

Figure 4.4: Average estimated initial dislocation-density in each ferrite grain from the full-field RVE results compared to the martensite grain boundary coverage c_{FM}. The dislocation density is dependent on grain size and relative martensite grain-boundary coverage. The bubble diameter is proportional to the equivalent grain size, the data is separated at 33 % and 66 % of the total number of ferrite grains. Solid lines represent least-squares fits for the three different respective grain-size groups.

initial average dislocation-density inside ferrite grains. The fixed width of the finite GND layer gives rise to a size-dependent average initial dislocation-density in ferrite grains. Therefore, the relative martensite coverage is considered to be a structural parameter of the DP microstructure. In order to quantify this martensite coverage, the coefficient c_{FM} is used. Based on the work of Calcagnotto et al. (2010a), the relative martensite coverage c_{FM} is defined by the phase boundary covered by martensite, l_{FM}, with respect to the total grain-boundary length, $l_{\mathrm{FF}} + l_{\mathrm{FM}}$,

$$c_{\mathrm{FM}} = \frac{l_{\mathrm{FM}}}{l_{\mathrm{FF}} + l_{\mathrm{FM}}}. \tag{4.25}$$

With this coefficient, the influence of the entire spatial arrangement of martensite is reduced to one scalar parameter.

Figure 4.4 depicts the difference in the average dislocation-density (estimated by Equation (4.23)) in each ferrite grain over the relative martensite-coverage c_{FM}. The data is colorcoded by grain-size intervals, and the diameter of the depicted bubbles is proportional to the EBSD grain-area. For each grain-size interval, a linear least-squares fit is shown in the evaluation figures.

There is a clear correlation apparent. An increased average initial dislocation-density coincides with a higher martensite-coverage in conjunction with the decreasing grain-size in ferrite. For smaller grain sizes, the misorientation at ferrite-ferrite interfaces influences a higher relative amount of the grain. In other words, the relative amount of the grain-interface with respect to grain-volume ratio increases (grain-interface to grain-volume ratio). This results in high scatter of the dislocation density in small ferrite grains, because the grain has to accommodate for the incompatibility within a smaller region. Small ferrite grains can show substantial work-hardening if they are surrounded by sufficiently high amounts of ferrite-martensite interfaces, or if they are small enough to be completely covered by a

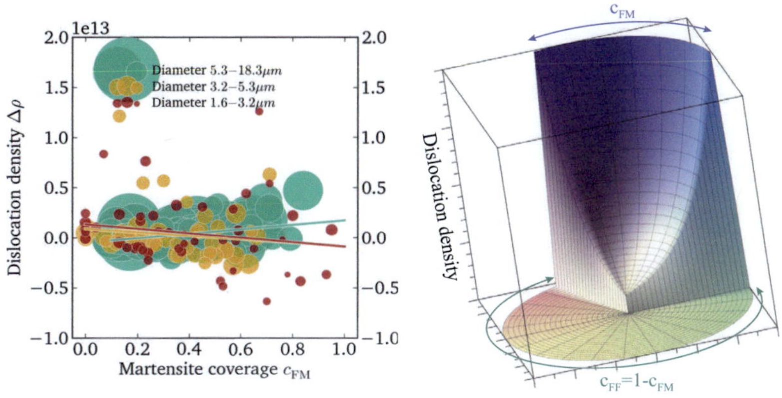

(a) (b)

Figure 4.5: Difference of the initial dislocation-density ρ_0 (Figure 4.4) to the model predictions from Equation (4.27) (a). The model prediction shows a good correlation (difference close to zero) for different grain sizes and martensite coverages. The bubble diameter is proportional to the equivalent grain-size, and the data is separated at 33 % and 66 % of the total number of ferrite grains. Different grain-size groups show similar results. The model grain for the initial dislocation density estimation is shown in Figure (b).

GND layer. With increasing grain size, ferrite grains become mechanically less influenced by surrounding martensite because the ratio of grain-interface to grain-volume decreases. An increasing interface-to-area ratio increases the initial deformation and, thus, the average dislocation-density of DP steels (Calcagnotto et al., 2010a). Experimentally, high initial dislocation-densities correlate with increasing martensite volume-fractions. This correlation is commonly explained by the additional contributions of GNDs caused by the ferrite-martensite interfaces.

The crystallographic orientations in ferrite grains (as modeled by crystal plasticity) cause complex interactions at grain- and phase-boundaries. This leads to increased fluctuations compared to purely isotropic calculations (Choi et al., 2013). The simulated dislocation-density distribution shows more obvious trend for the martensite coverage when compared to the experimentally measured distribution in Calcagnotto et al. (2010a). This is mainly due to the absence of three-dimensional spatial arrangements in this simulation. For the initial state, it is simple to isolate the martensite influence on ferrite from additional effects. In the simulations the pre-quenching dislocation-density is uniformly distributed as opposed to experiments. Additionally, the effect of martensite islands is restricted to regions close to ferrite-martensite-interfaces.

4.4 Initial dislocation-density approximation

Based on the individual grain-boundary ratio of each ferrite grain that is covered by martensite c_{FM} (Section 4.3), the initial dislocation density can be separately approximated. The proposed approximation makes use of a simplified two-dimensional grain, which is composed of two different regions as depicted in Figure 4.5(b). Influenced by the austenite-martensite phase transformation, the martensite-covered region shows a certain gradient in the dislocation-density. The exponential part of the assumed piecewise defined function

$$\rho(r,\phi) = \begin{cases} \rho_{\mathrm{GB}} - (\rho_{\mathrm{GB}} - \rho_{\mathrm{GI}})\left(1 - \exp\left(-\tilde{\lambda} r\right)\right) & \text{for } \phi \in (0, c_{\mathrm{FM}} \cdot 2\pi) \\ \rho_{\mathrm{GI}} = \text{const.} > 0 & \text{for } \phi \in (c_{\mathrm{FM}} \cdot 2\pi, 2\pi) \end{cases} \tag{4.26}$$

was fitted to match the overall initial dislocation-density distribution from Figure 4.4 and the line scan profile of Figure 4.8 ($\rho_{\mathrm{GB}} = 2 \cdot 10^{13}\,\mathrm{m}^{-2}$ and $\tilde{\lambda} = 1.254$). Near ferrite-ferrite interfaces, the second region in Figure 4.5(b) shows no initial inhomogeneities from the austenite-martensite phase transformation. It has a comparably low and constant dislocation-density of $\rho_{\mathrm{GI}} = 1 \cdot 10^{11}\,\mathrm{m}^{-2}$, as found in the grain interior. This allows for an analytical result for the average dislocation density in the model grain

$$\bar{\rho}_0(D, c_{\mathrm{FM}}) = \rho_{\mathrm{GI}} + (\rho_{\mathrm{GB}} - \rho_{\mathrm{GI}}) \frac{4c_{\mathrm{FM}}}{\tilde{\lambda}^2 D^2} \left[2 \exp\left(-\frac{\tilde{\lambda} D}{2} \right) - 1 + \tilde{\lambda} D \right] \quad (4.27)$$

The martensite coverage c_{FM} of each ferrite grain was extracted from the EBSD data in Figure 4.1. Figure 4.5(a) shows least-squares fits of the difference between the model prediction (Equation (4.27)) and the simulation data for each ferrite grain from Figure 4.4 (for an ideally matching model, the difference would be zero in all grains). For the presented model, the least-square fits show only small deviations from the simulation results for different grain sizes. The model reasonably predicts the grain-size dependent initial dislocation density in the RVE since the overall difference is close to zero. The simplicity of the model certainly limits its prediction accuracy. For example, highly non-spherical grains or martensite inclusions which are completely embedded inside ferrite grains, are incorrectly approximated.

4.5 Microscopic material state evolution

In general, martensite is subjected to higher average loads compared to the ferrite matrix in DP steels (Figure 4.6), while the ferritic matrix carries the majority of the deformation strain. The resulting stress concentration in martensite islands mainly stems from the pronounced composite effect due to a yield-strength ratio of $2 - 4$ in DP steels. Circular martensite-islands are inhomogeneously loaded (Figure 4.6) and thus subjected to lower average stresses compared to banded martensite islands (Tasan et al., 2010; Azuma et al., 2012). It is also evident that some of the initial regions of low deformation remain unhardened up to high macroscopic loads (white arrows in Figure 4.6). These areas

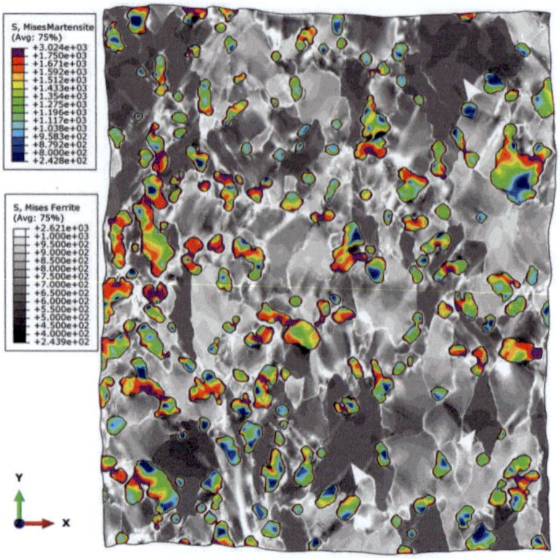

Figure 4.6: Equivalent von Mises stress in the RVE with quenching after $20\,\%$ macroscopic compression strain. Two different spectra are used: colored for martensite, and black/white for ferrite. Martensite is inhomogeneously loaded. Circular martensite inclusions are only loaded near the ferrite-martensite grain boundary. Elongated martensite inclusions, however, are subjected to a bending-like load and exhibit higher stresses. Most ferrite-ferrite grain boundaries are still recognizable despite the overall high dislocation density.

coincide with areas without martensite particles, and indicate unused work-hardening capacity of the DP steel (Tasan et al., 2014).

At ferrite-ferrite interfaces, crystallographic misorientation leads to stress-strain incompatibilities and the evolution of shear bands is severely affected. While a shear band might be favorable in one ferrite grain, the next grain can have a crystallographic orientation that seriously inhibits the shear-band propagation. The influence of crystallographic orientation on the average mechanical behavior is most pronounced in small grains. This is cause by the high interface-to-area (grain boundary to grain area) ratio (Calcagnotto et al., 2010a).

In crystal-plasticity simulations, ferrite-ferrite interfaces have a distinct effect on the dislocation-density distribution due to their inher-

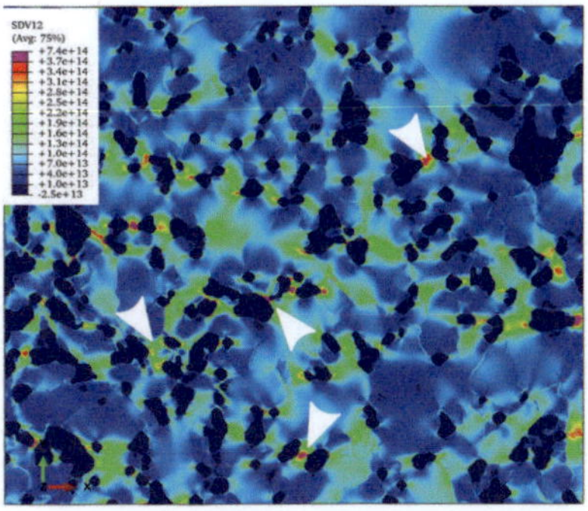

(a) at $\bar{\varepsilon} = 2\,\%$

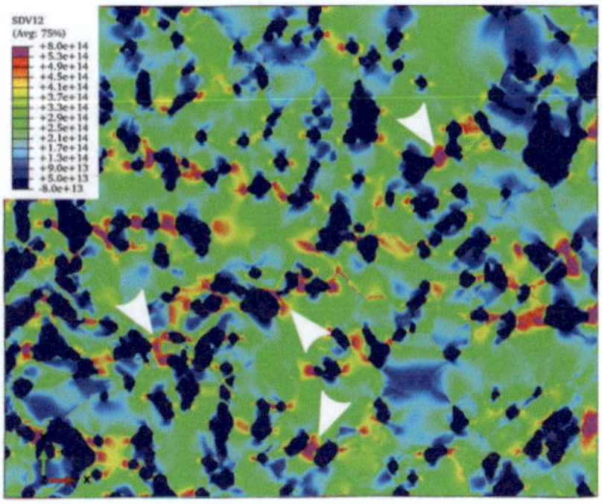

(b) at $\bar{\varepsilon} = 5\,\%$

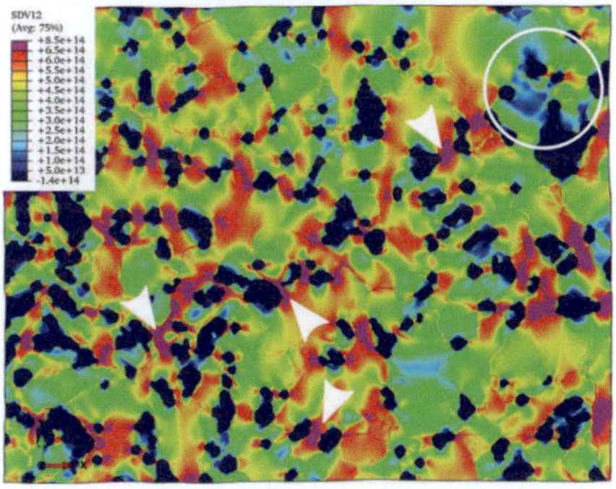

(c) at $\bar{\varepsilon} = 10\,\%$

Figure 4.7: Estimated dislocation density in the full-field RVE with quenching after increasing macroscopic strain $2\,\%$ (a), $5\,\%$ (b), and $10\,\%$ (c). As it is experimentally observed, the dislocation-density distribution shows an interrupted and inhomogeneous pattern due to the present crystallographic misorientation. Localization and failure hotspots are highlighted and a low deformation area is encircled.

ent incompatibility. Thus, the general stress level in the RVE of this work shows higher fluctuations when compared to isotropic calculations of, e.g., Sodjit and Uthaisangsuk (2012) or Ramazani et al. (2013). Choi et al. (2013) found much more scattered stresses in ferrite and martensite for the crystal-plasticity simulation. The aforementioned effect on shear-band evolution is observed, too. The irregular shear bands observed in crystal-plasticity simulations are closer to the experimental results. This can be confirmed by a visual comparison between the crystal-plasticity simulations and the experimental works of Kapp et al. (2011) and Tasan et al. (2010), respectively. In this works, crystal-plasticity finite-element simulations predict a correct overall strain pattern but are unable to resolve the highly-local and sharp strain patterns evident in experimental measurements (Tasan et al., 2014). Nevertheless, without significant local martensite heterogeneities, it is experimentally observed that plastic deformation starts in the largest ferrite-grains in DP steels. Therefore, in the work of Tasan et al. (2014), grain-size is considered to have a higher impact on deformation localization compared to the present crystallographic orientation.

Even for low macroscopic deformation, high deformation-fluctuations are present in ferrite (Figure 4.7). The initial strain-localization hot-spots show almost unchanged locations, but the localization accelerates with increasing macroscopic deformation. Relatively isolated and interrupted shear bands in ferrite grains are present, nevertheless. At the same time, however, some ferrite regions remain at a very low relative strain-ratio of only one-third of the macroscopic strain, even at high macroscopic strains.

The qualitative findings are in line with the experimental observations in Kapp et al. (2011). However, strain partitioning is more pronounced for higher macroscopic strains in their work. The RVE clearly shows localized strain-areas after about 5 % macroscopic tensile strain. Then DP steel modeled in the RVE begins to fail because material hardening is not sufficient to inhibit the propagation of areas of failure. As depicted in Figure 4.7, localization hot-spots are invariably located near martensite islands. The earliest localizations, additionally, lie within a shear band. Ferrite shear-bands are channeled by martensite into small ferrite passages and, therefore, lead to failure.

Failure. At 10 % macroscopic tensile strain, failure-initiating hot-spots can be clearly identified. It is reasoned that the decrease of the global work-hardening rate (which will be discussed later based on Figure 4.8 and Figure 4.11) is a combination of the work-hardening decrease of the ferrite phase and localizations inside the RVE. A tensile-failure strain of $\sim$ 12 % is much lower than it is experimentally observed for most DP600 steels. Furthermore, it was experimentally found, that damage events are primarily located at the borders of high-deformation zones that often lie within large ferrite grains (Tasan et al., 2014). The numerically determined low tensile-failure strain is a clear limitation of the conducted simulation because the early localization for tensile loading could not be prevented for reasonable material parameters. To circumvent the influence of premature localization, compression simulations were carried out to evaluate the work-hardening behavior at higher strains.

4.5.1 Spatial dislocation-density evolution

Figure 4.8 is a „text-book" example that fits the common understanding of dislocation-density evolution processes in DP steels. It will be discussed in more detail that the general trends are similar to this example albeit high fluctuations are possible for different line-scan paths. Figure 4.8 depicts the evolution of the estimated dislocation-density along the line scan indicated in Figure 4.3. Starting at a ferrite-martensite interface on the left end, the line scan crosses a ferrite-ferrite interface and ends at a ferrite-martensite interface again.

The aforementioned gradient, from high dislocation-densities at ferrite-martensite interfaces to low densities in the grain interior, is present after the cooling step at 0 %. Such a gradient is absent at ferrite-ferrite interfaces as can be observed in Figure 4.8(a). With increasing macroscopic strain, the dislocation-density gradient shifts towards the grain interior. Then, the ferrite-ferrite interface becomes detectable

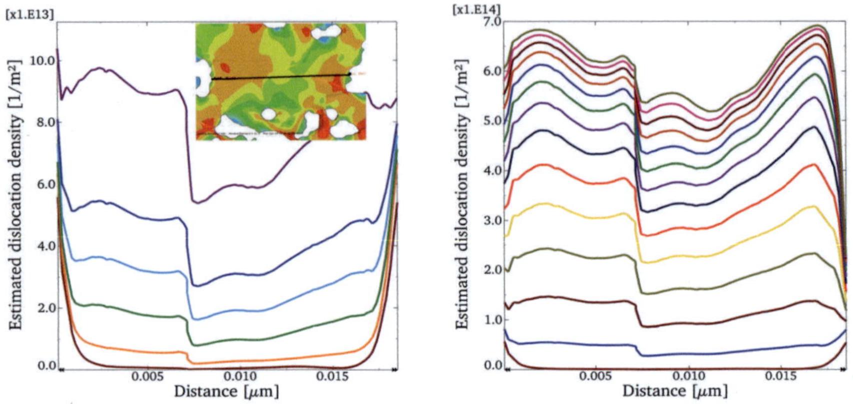

(a) (b)

Figure 4.8: Monotonically increasing evolution of the dislocation density along the line scan at $0\,\%, 0.25\,\%, 0.5\,\%, 0.75\,\%, 1\,\%, 1.5\,\%$ macroscopic strain (a); and monotonically increasing in $1\,\%$ percent steps from $0\,\%$ up to $13\,\%$ macroscopic strain (b). Figure (a) and (b) use different scales, but the lowest curves are identical. The dislocation density distribution shows the characteristic gradient near ferrite-martensite phase-boundaries after quenching (Figure a). The dislocation pattern remains self similar after $\sim 1\,\%$ macroscopic deformation (Figure b). Martensite boundaries appear to shield adjacent ferrite regions from deformation. A conclusion drawn from the lower dislocation densities on the start and end of the line scan (b).

as a jump in the dislocation density. The dislocation-density evolution in these early macroscopic deformation-stages is determined by inhomogeneities from the initially graded GND-layer. Additionally, incompatibilities at ferrite-ferrite interfaces facilitate an increased average dislocation-density storage.

Starting from approximately $0.5\,\%$ macroscopic strain, the gradient evolution transitions into a state where a particular dislocation-density fluctuation-pattern develops. With increasing macroscopic deformation, „regular" work-hardening of ferrite becomes dominant and the initial inhomogeneity from the GND layer is not identifiable any more. Sarosiek and Owen (1984) suggest a similar explanation for the work-hardening in DP steels. They attribute the initially high work-hardening to forest hardening from heterogeneously deforming ferrite, i.e., the degradation of the initial gradient into the pristine interior. The lower work-hardening at higher strains is attributed to forest hardening with a uniform dislocation density in their work. Here, it is analogously explained by the absent change of the work-hardening fluctuation-pattern.

After this transition, the dislocation-density fluctuation-pattern remains self-similar but the dislocation-density is shifted to higher values (Figure 4.8b). This is in line with experimental results from Kapp et al. (2011). They found that the locations of the initial strain-localization hot-spots remain stationary but scale with increasing macroscopic deformation (see also Tasan et al., 2014). The average fluctuation-range within the grain is about an order of magnitude smaller than the dislocation-density produced by ferrite hardening. Therefore, a description of the work-hardening solely based on the average dislocation-density (as done in Chapter 6) appears to be a justifiable simplification.

The experimentally observed ferrite dislocation-density becomes very high, and the sharp contrast at ferrite-ferrite interfaces vanishes. Consequently, the distinction of individual ferrite grains becomes increasingly difficult at higher macroscopic strains (Korzekwa et al., 1984; Dillien et al., 2010; Sarosiek and Owen, 1984). This effect can be recognized in the RVE results. For low deformations, most ferrite-ferrite grain boundaries are easy to determine in Figure 4.7; for large deformations, shear bands spanning over ferrite-ferrite interfaces without a visible jump in the dislocation-density are present, too (Figure 4.7c).

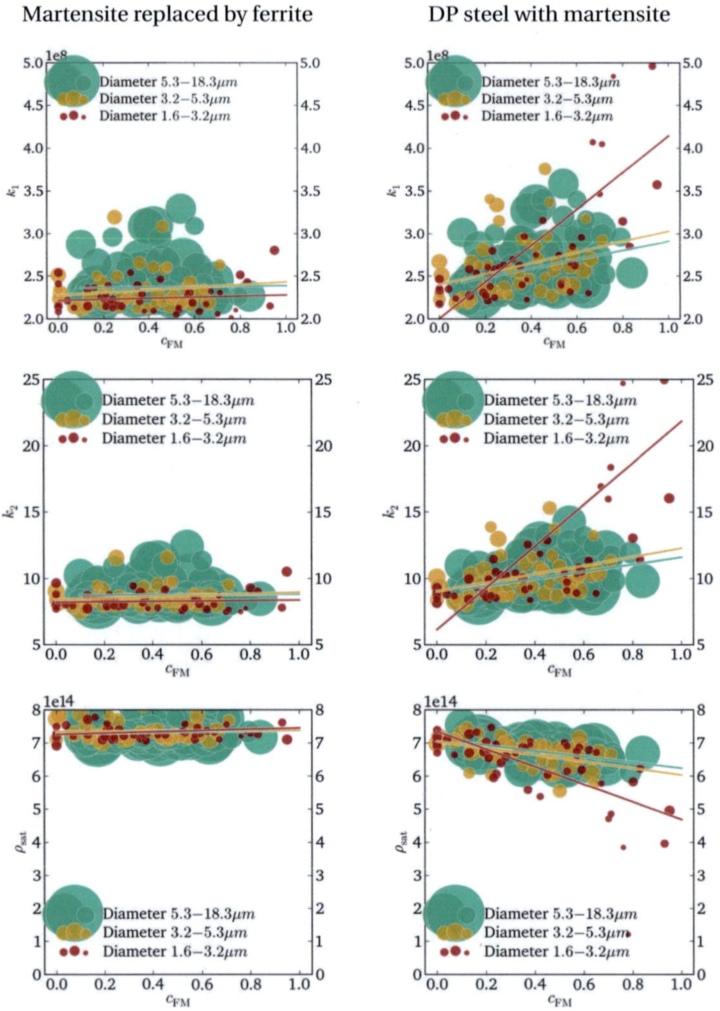

Figure 4.9: Results from the full-field RVEs with martensite (right column) and with only ferrite (left column): dislocation production parameter k_1 (top row); dislocation-recovery parameter k_2 (middle row); saturated dislocation-density ρ_{sat} (bottom row). The bubble diameter is proportional to the equivalent grain size, the data is separated at $33\,\%$ and $66\,\%$ of the total number of ferrite grains. Solid lines represent least-square fits for the respective grain-size groups.

4.5.2 Statistical evaluation of work-hardening parameters

So far, the analysis of the micromechanical behavior in DP steels has been limited to visual observations and the interpretation of local regions of interest. For a more fundamental understanding of the actual effect of the martensite inclusions´ presence on ferrite work-hardening, a comparison simulation is conducted (Figure 4.9). In this simulation, all martensite inclusions are replaced by an isotropic phase with the average ferrite work-hardening evolution. This procedure allows for a statistical grain-by-grain comparison (Figure 4.10). Hence, it enables separation of the martensite influence from other effects causing deviations to work hardening. For example, grain orientation and incompatibility at ferrite-ferrite interfaces or the spatial arrangement of the ferrite grains are present in DP-steels, too.

Figure 4.10 depicts the difference in the two work-hardening parameters dislocation-density storage k_1 (Figure 4.10a), and saturated dislocation-density, $\rho_{\mathrm{sat}} = k_1^2 k_2^{-2}$ (Figure 4.10b). The parameters were found by a least-squares fit of Equation (6.2) to each individual ferrite grain. Note that the Kocks-Mecking-Estrin model is equivalent to the Voce model of the work-hardening in the crystal-plasticity simulation. The difference $\Delta\psi = \psi_{\mathrm{withM}} - \psi_{\mathrm{withoutM}}$ was computed for each ferrite grain. The quantity simulated with actual martensite (ψ_{withM}) is taken as reference. From this reference the respective quantity simulated with martensite replaced by ferrite (ψ_{withoutM}) is subtracted. All values are plotted over the relative martensite coverage c_{FM}.

Surprisingly, the average effect of martensite coverage reverses during the loading process. At the beginning of the deformation, a higher martensite-coverage is linked to increased dislocation-densities (Figure 4.4) and increased dislocation-storage in ferrite grains (Figure 4.10a). With ongoing deformation, however, martensite seems to prohibit small ferrite-grains from work hardening and the saturated

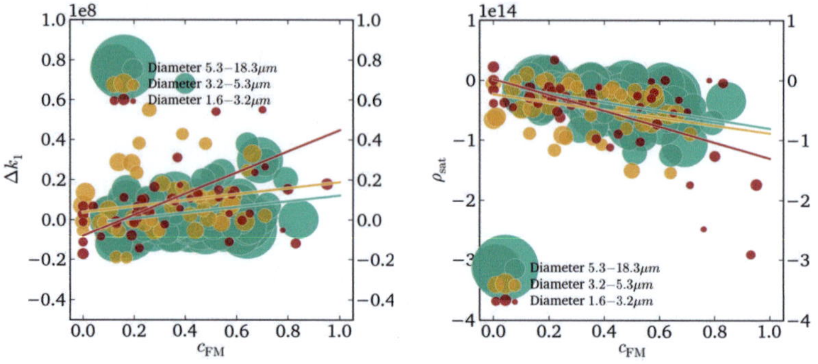

(a) (b)

Figure 4.10: Per-grain difference of the average dislocation production parameter k_1 between the full-field simulations with martensite and with ferrite only (a); per-grain difference of the saturated dislocation density ρ_{sat} (c). Both quantities are dependent on grain size and relative martensite grain-boundary coverage. However, the effect of martensite on the dislocation density is inverse for the saturated state (b) when compared to the initial state of Figure 4.4. The bubble diameter is proportional to the equivalent grain size, the data is separated at 33 % and 66 % of the total number of ferrite grains. Solid lines represent least-square fits for the respective grain-size groups. The original data for both RVEs can be found in Figure 4.9.

dislocation-density decreases (Figure 4.10b). Since $\rho_{\mathrm{sat}} = k_1^2 k_2^{-2}$, this is equivalent to stating that with increasing martensite-coverage k_2 increases more than k_1. At the right end of the line scan in Figure 4.8(b) martensite shields an area from deformation as can be seen in the inset. Again, a grain-size effect is notable. Jiang et al. (1995) also reported a similar trend for DP steels: an increasing martensite volume-fraction leads to an increased work-hardening rate at low strains but significantly decreased rates at higher strains. Nesterova et al. (2015) found a delayed dislocation-patterning in small grains with a high relative martensite-coverage. Nevertheless, high gradients were present in these grains, both in the simulation and experimental results.

Nano-indentation experiments from the work of Ghassemi-Armaki et al. (2014) support the observations from Figure 4.10. In their work, ferrite located within a range of $\sim$ 3 µm near ferrite-martensite interfaces, exhibits two opposing effects. After quenching, these regions show a higher microhardness compared to the grain interior due to work hardening from the phase transformation. However, for a significant macroscopic deformation of $\sim$ 7 %, the same regions near ferrite-martensite interfaces exhibit a lower microhardness. The microhardness is lower compared to the ferrite interior, and even compared to the initial hardness of these regions. Nevertheless, Ghassemi-Armaki et al. (2014) observe overall work-hardening for the averaged behavior of ferrite grains, as it would commonly be expected.

This experimentally observed change of the influence of ferrite-martensite interfaces on the deformation of ferrite is similar to the effect found here. Ghassemi-Armaki et al. (2014) explain the softening effect near ferrite-martensite interfaces by complex interactions of GNDs and statistically-stored dislocations. However, the crystal-plasticity hardening model of the work at hand does not cover these interactions. From these shortcomings model, it can be deduced that the experimentally observed behavior can be, at least in part, attributed to the purely-spatial effect that was observed, here.

The work of Bergström et al. (2010) also supports the findings of the work at hand. They find that a higher martensite content and higher martensite shielding decreases the amount of actively hardened ferrite areas in DP steels. If necking occurs, ferrite is almost completely deformed in the DP500 steel. In this case, only $\sim$ 2% of the material remains undeformed. Contrarily, in the DP980 steel almost 20% of the ferrite remains undeformed after necking.

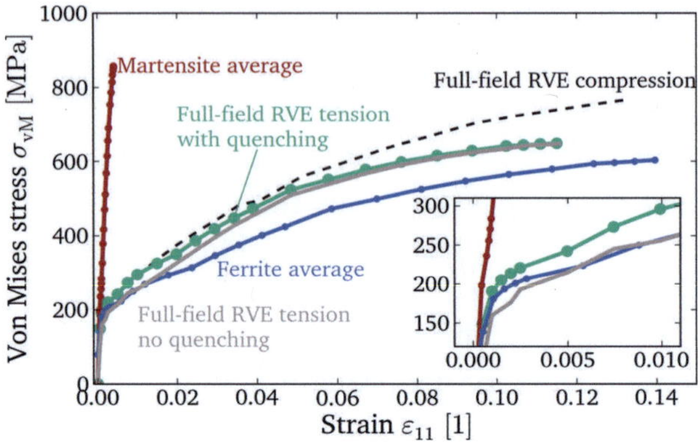

Figure 4.11: Macroscopic stress-strain behavior for the simulated full-field RVEs: tensile load after quenching and without quenching (solid lines); and compression load (dashed). The influence of the initial compatibility deformation after quenching is notable up to $\sim 8\,\%$.

4.6 RVE based macroscopic behavior prediction

Figure 4.11 shows the macroscopic behavior of the simulated RVE with and, respectively, without initial GNDs from the austenite-martensite phase transformation during quenching. The RVE calculation with quenching shows a slightly increased yield-strength. As discussed here, this strengthening can be explained by an increased dislocation-density, that is also experimentally observed (Korzekwa et al., 1984; Ramazani et al., 2013). The higher initial work-hardening rate and the rapid decrease of the work-hardening rate are also observed in experiments (Korzekwa et al., 1984). After $\sim 8\,\%$ tensile deformation, the difference between the two simulations vanishes entirely, thus, highlighting the increasing amount of plastic-deformation compared to the initial GND-gradient. A comparison between the two curves for tension and compression reveals loading-asymmetry and localization, starting as early as from $\sim 4\,\%$ strain.

The simulated RVE exhibits very heterogeneous strain-partitioning. From Figure 4.11 it is concluded that the vast majority of martensite stays elastic at $\sim 10\,\%$ strain, while ferrite carries almost the entire deformation strain. Kang et al. (2007) found a maximum strain-partitioning at discrete points of up to $\varepsilon^{\mathrm{F}}/\varepsilon^{\mathrm{M}} \simeq 18$, while the global partitioning in this work is increasing up to $\varepsilon^{\mathrm{F}}/\varepsilon^{\mathrm{M}} \simeq 30$. In view of the discussion in Chapter 6, the artificially high strain-partitioning can be attributed to the lack of non-local effects in the material models. For example, full-field simulations, commonly, do not account for long-range stresses of dislocation patterns.

Chapter 5

Experimental methods

The experimental data documented in this chapter has been measured and post-processed by several co-workers from the Karlsruhe Institute of Technology (KIT)[1]. The material under consideration is a DP600 steel from Salzgitter Flachstahl GmbH, Germany:

- Section 5.1: René Sebastiano under supervision of Barthel Brylka, Institute of Engineering Mechanics (ITM) – Chair for Continuum Mechanics (Sebastiano and Brylka, 2014; Sebastiano, 2015)
- Section 5.2: Moritz Wenk, Institute for Applied Materials – Materials-mechanics and Biomechanics (IAM-WBM), DFG-GRK1483 Subproject A11 (Wenk, 2014)
- Section 5.3: Simone Schuster, Institute for Applied Materials – Materials Science (IAM-WK), DFG-GRK1483 Subproject A13 (Schuster, 2014)

The fundamental information about the experimental setup and methodology from these works is repeated in this section for the sake of completeness.

[1] This section is reproduced from the submitted article: F. Rieger, T. Böhlke, Mechanism based mean-field modeling of the work-hardening behavior of dual-phase steels.

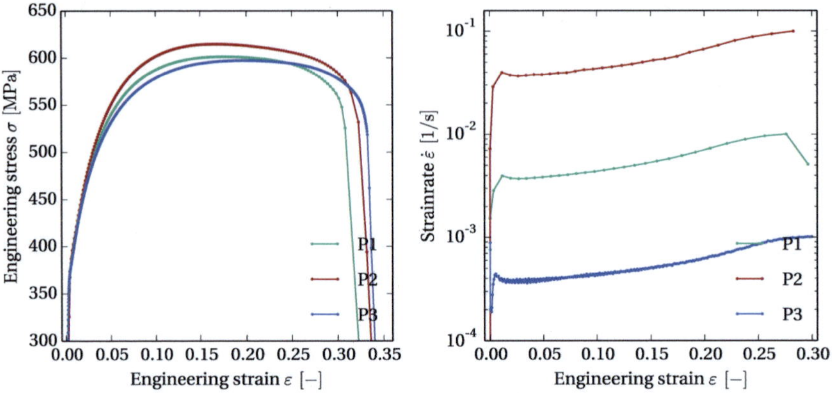

Figure 5.1: Uniaxial tension of the investigated DP600 steel for strain rates (right) of about $\dot{\varepsilon} \simeq 10^{-3}, 10^{-2}, 10^{-1}\,\mathrm{s}^{-1}$ (Sebastiano, 2015, Fig. 4.5). The investigated strain-rates show only minor influence on the overall mechanical behavior. Nevertheless, the material model can be easily extended to incorporate a simple strain-rate dependence (Section 6.1).

5.1 Material characterization

Uniaxial and biaxial tensile experiments documented in this paragraph were conducted by Rene Sebastiano (Sebastiano and Brylka, 2014; Sebastiano, 2015). The testing was done with a biaxial-tension device manufactured by Zwick/Roell with 150 kN load cells (Zwick GmbH & Co. KG, Ulm, Germany; single-unit production, property of the Institute of Engineering Mechanics (ITM) – Chair for Continuum Mechanics). Strain data was recorded with a built-in optical extensiometer. The full-field strain measurements were conducted with an optical GOM 3D-measurement system (Gesellschaft für Optische Messtechnik mbH, Braunschweig, Germany). Digital image-correlation has been carried out with the commercial software ARAMIS (GOM). Additionally, an in-house code was used as well (Sebastiano and Brylka, 2014; Sebastiano, 2015).

Uniaxial-tension. Uniaxial-tension experiments (Figure 5.1 and Figure 7.2a) were conducted according to the NORM DIN EN ISO 6892-1 (2014). The overall results exhibit a small scatter for repeated experiments. The influence of different sheet orientations $(0°, 45°, 90°)$ is small, as discussed below. Results from the built-in optical extensiometer and the ARAMIS-software coincide for this basic experiment. The DP600 steel exhibits a slightly higher yield strength with increasing strain-rate in the investigated strain-rate regime (Figure 5.1). This is not incorporated in the model but will be elaborated on in Section 6.1.

Quasi-isotropy of DP600. For practical applications, the amount of plastic anisotropy in sheet metals is often measured with the Lankford coefficients. Lankford coefficients, or r-values, are defined as

$$r_\theta := \frac{\varepsilon_{22}^{\mathrm{p}}(\theta)}{\varepsilon_{33}^{\mathrm{p}}}. \tag{5.1}$$

Equation (5.1) computes the ratio between the strains in width direction, $\varepsilon_{22}^{\mathrm{p}}$, with respect to strains in thickness direction, $\varepsilon_{33}^{\mathrm{p}}$. Sheet metals exhibit two different types of anisotropy, namely planar and normal anisotropy (Gambin, 2002).

Normal anisotropy describes the difference between (averaged) plastic properties in the directions within the sheet plane and in thickness direction. It is measured by the average strain-ratio given by

$$r_{\mathrm{m}} = \frac{1}{4}\left(r_{0°} + 2r_{45°} + r_{90°}\right). \tag{5.2}$$

Normal anisotropy is often taken as a measure for the drawability of a sheet steel as it is related to the resistance against local thinning. High values of $r_{\mathrm{m}} \gtrsim 1$ indicate good deep-drawing properties. The investigated DP600 steel shows reasonable drawability with $r_{\mathrm{m}} = 0.93$; for an isotropic material one finds $r_{\mathrm{m}} = 1$.

The plane-strain ratio Δr is used as a measure for planar anisotropy,

$$\Delta r = \frac{1}{2}\left(r_{0^\circ} - 2r_{45^\circ} + r_{90^\circ}\right). \tag{5.3}$$

It describes the difference of plastic properties in the directions within the sheet plane. For the investigated DP600 steel, the plain strain ratio is $\Delta r = 0.075$. In this context the DP steel shows a comparably close behavior to a planar-isotropic material which would exhibit a plane-strain ratio of $\Delta r = 0$.

Earing profile. During cup-drawing experiments with metals, one commonly observes earing profiles. The number of ears and the earing depth is mainly controlled by the planar anisotropy. The investigated DP600 steel exhibits a total of four ears. According to Gambin (2002), the quadratic Hill-Criterion is able to model a four-ear profile, since it fully accounts for orthotropic material symmetry. Orthotropic materials have mutually-orthogonal symmetry planes in rolling, transverse and normal direction (Barlat and Lian, 1989).

Yoon et al. (2006) highlight the direct influence of planar anisotropy on the earing profile. Consequently, they proposed a model to estimate the earing profile directly from the measured r-values by

$$\Delta h\left(\theta\right) = \frac{r_{(\theta+90^\circ)}}{r_{(\theta+90^\circ)} + 1}\left(R_\mathrm{c} - R_\mathrm{b} + R_\mathrm{b}\ln\left(\frac{R_\mathrm{b}}{R_\mathrm{c}}\right)\right). \tag{5.4}$$

Using the blank and cup diameters ($R_\mathrm{b} = 200\,\mathrm{mm}$ and $R_\mathrm{c} = 100\,\mathrm{mm}$) one finds a low earing-profile amplitude of $1.7\,\mathrm{mm}$ or 1.7% for the DP600 steel. This is close to the experimental results of $\sim 2\,\mathrm{mm}$ (measured by the author of the work at hand). From the r-values of Table 5.1, one would expect only two ears using the formula of Yoon et al. (2006) (at 0° and 180°) since the r-values between 0° and 90° have only one

Angle	$0°$	$45°$	$90°$
r-value	0.88	0.89	1.05

Table 5.1: Lankford r-values of the DP600 steel for different angels (Sebastiano, 2015, Tab. 4.4). The overall anisotropy of the DP600 steel is small (planar anisotropy $\Delta r = 0.075$, normal anisotropy $r_\mathrm{m} = 0.93$) which is confirmed by the low earing-profile height of about $1.7\,\%$ of the cup height.

maximum at $90°$. Cups from conducted deep-drawing experiments, however, exhibit four ears.

Monotonic biaxial-tension. Monotonic biaxial-tension experiments were conducted according to the NORM DIN EN ISO 16842 (2014); cross-shaped specimen with grooved arms (Figure 5.2a) are biaxially loaded with different load ratios. Because DP steels commonly exhibit continuous yielding, the yield surface for monotonic loading was calculated from the "contours of equivalent plastic-work" (Hill et al., 1994). Both the isotropic von Mises and the anisotropic Mises-Hill yield surface agree reasonably with the measured data due to the DP600 steel being only slightly anisotropic (Figure 5.2b). For higher loads, however, long-range stresses shift the origin and distortional hardening deforms the shape of the yield-surface (e.g., Barlat et al., 2013). Distortional hardening is not considered, here. The isotropic von Mises criterion is used in this work because the anisotropy is rather small for the investigated DP600 steel and for DP steels in general (Resende and Bouvier, 2013). For example, Tarigopula et al. (2009) found the that this criterion gives reasonable results for DP steels, too.

Subsequent-$90°$ tension. To identify the kinematic-adjustment rate (the material parameter c, see Section 6.4.2), a loading sequence called "subsequent-$90°$ tension" is conducted. The sequence consists of loading along the x-direction, then unloading, and finally loading along the y-direction (for details, see the discussion in Sebastiano (2015)). In or-

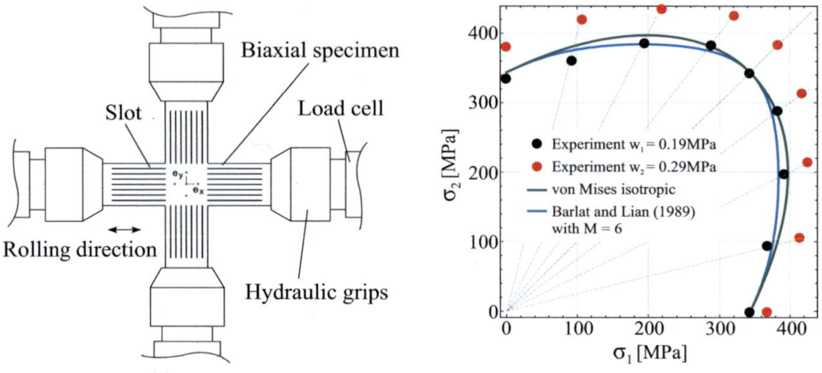

(a) (b)

Figure 5.2: Specimen geometry and experimental set-up (a, Sebastiano (2015, Fig. 3.1)); monotonic biaxial-loading with different load ratios for DP600 steel (b, Sebastiano (2015, Fig. 4.28)). Dots show experimental data for increasing plastic work; w_1 is equivalent to 344 MPa for uniaxial tension. The initial von Mises (green) and the Barlat and Lian (1989) yield surface (blue) are shown. Both agree reasonably with the yield surface for w_1, since the DP600 steel is only slightly anisotropic.

der to quantify such abrupt strain-path changes, the Schmitt factor, given by (Schmitt et al., 1985, 1994)

$$\theta = \frac{\boldsymbol{\varepsilon}^{\mathrm{p}}_{\mathrm{pre}} \cdot \boldsymbol{\varepsilon}^{\mathrm{p}}_{\mathrm{sub}}}{\|\boldsymbol{\varepsilon}^{\mathrm{p}}_{\mathrm{pre}}\| \, \|\boldsymbol{\varepsilon}^{\mathrm{p}}_{\mathrm{sub}}\|}, \tag{5.5}$$

is commonly used as a scalar indicator (Rauch et al., 2011). It essentially computes the scalar product (or angle) between the plastic prestrain $\varepsilon^{\mathrm{p}}_{\mathrm{pre}}$ and subsequent plastic loading $\varepsilon^{\mathrm{p}}_{\mathrm{sub}}$. For reloading in the same direction one finds $\theta = 1$, for exact strain-reversal or Bauschinger loading $\theta = -1$, for orthogonal-loading $\theta = 0$ and for the employed subsequent-$90°$ tension sequence $\theta = -0.5$ (Rauch et al., 2011). Orthogonal-loading can be achieved by, e.g., tension and subsequent simple shear in the same direction. Alternatively, one may consider shear in one direction and subsequent shear with an angle of

90° to the original direction. This loading sequence is close to the case of cross-loading. During cross-loading, all activated slip-systems from pre-straining are inactive during the subsequent loading (Rauch et al., 2011). Cross-loading, essentially, exhibits the maximal latent hardening effect of all strain-path changes; almost unchanged pre-straining dislocation-structures interact with the newly activated slip-systems (Barlat et al., 2013). The subsequent-90° tension lies in between cross-loading and reverse loading and, therefore, should exhibit a behavior in between these two extremal loading cases.

Figure 5.3 depicts the post-processed stress over the equivalent plastic strain. There is a significant Bauschinger or early re-yielding effect. No hardening plateau due to cross-hardening or softening is observed. Therefore, and due to the inherent complexity, cross-hardening effects are not considered in this work. Additionally, strain-path changes during deep-drawing are continuous. Therefore, the strain-path change indicator should use an evolving representative measure of the current plastic strain. Comparison of the instantaneous values for θ (Equation (5.5)) would lead to insensitivity to continuous changes. In the case of oscillations around an average value, instantaneous values for θ result in an overestimation of the influence of the strain-path change (Carvalho-Resende et al., 2013).

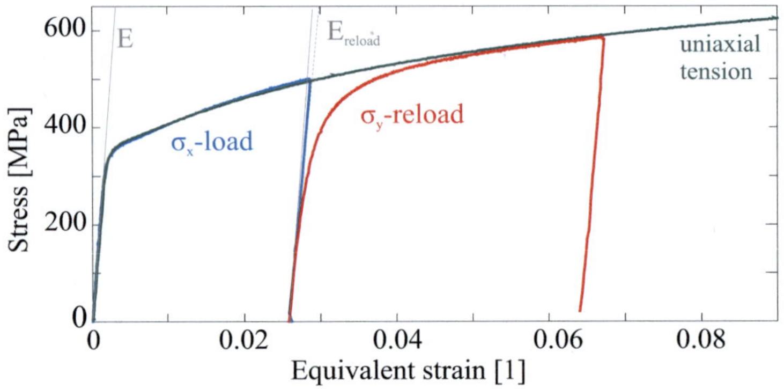

Figure 5.3: Subsequent 90°-tensile loading for DP600 steel (along the x-direction (red), unload, then along the y-direction(green)); uniaxial tension loading is shown for comparison (Sebastiano, 2015, Fig. 4.36). There is a significant Bauschinger or early re-yielding effect, but no hardening plateau.

5.2 Microstructural investigations

Microstructural investigations were performed by Moritz Wenk as part of the Project A11 within the Graduate School 1483. [2)]The dual beam microscope (Nova Nanolab 200 FEI) has a SEM thermal field emission gun and a Ga^+ ion column, mounted at $52°$ against the electron column. Typical electron acceleration voltages for the carried out investigations are: $10\,\mathrm{kV}$ with a beam current of $0.54\,\mathrm{nA}$ for imaging; $20\,\mathrm{kV}$ and $6.6\,\mathrm{nA}$ for EBSD measurements with a charge-coupled-device phosphor screen. For milling micro-pillars with focused ion beams, an acceleration voltage of $30\,\mathrm{kV}$ and currents from $0.5\,\mathrm{nA}$ to $20\,\mathrm{nA}$ have been used (Schreijäg, 2013; Wenk (2014)).

EBSD measurements. The utilized EBSD detector was manufactured by HKL, now Oxford Instruments (Tubney Woods, Abington, Oxford-

[2)] Institute for Applied Materials – Materials- and Biomechanics (IAM-WBM) (Wenk and Schreijäg, 2013; Wenk, 2014)

C_{DP}	Si	Mn	C_F	C_M	**Reference**
0.15	0.3	1.45	0	0.33	DF140T, Chen et al. (2014)
0.09	0.6	2.15	0	0.15	DP980, Chen et al. (2014)
0.15	0.003	0.99	0	0.55	Park et al. (2014)
0.17	0.22	1.49	0.01	0.054	Calcagnotto et al. (2010b)
0.093	0.29	1.5	0.015	0.6	DP600, this work Schuster (2014)

Table 5.2: Chemical composition of the investigated DP600 DP steel (in wt.%) and DP steels from literature. The carbon contents of ferrite and martensite are estimations (Section 6.7).

shire, OX13 5QX, UK). The recorded Kikuchi patterns were analyzed using the "Channel 5" computer program from the same company. The DP600 specimen were first mechanically ground with SiC paper of decreasing grain size and then polished with diamond paste (6 μm, 3 μm and 1 μm). To remove deformations that were introduced by mechanical polishing, the samples were processed by vibratory polishing with oxide particles. Based on the misorientation ϑ, estimations of GND densities were calculated by $\rho_{\mathrm{GND}} = \vartheta / (ub)$ with respect to the step size u and the burgers vector b (Kubin and Mortensen, 2003; Schreijäg, 2013). Clearly, this estimation does not account for different slip systems and is a simplification.

Micro-pillars. The procedure for micro-pillar preparation consists of an initial coarse cut with high beam currents ($7 - 20\,\mathrm{nA}$) to remove most of the material around the pillar by multiple passes in circular patterns. A single last pass with low beam currents ($0.5-3\,\mathrm{nA}$) produces the final pillar. The resulting micro-pillars with diameters of 0.5, 1 and $2\,\mu\mathrm{m}$ exhibit a taper of $\sim 5°$. Pillars were cut from grains with $< 111 >$, $< 101 >$, $< 111 >$ and $< 123 >$ out-of-plane orientation-directions in ferrite. Martensite pillars with both a $2\,\mu\mathrm{m}$ diameter and a hight of $3\,\mu\mathrm{m}$ were difficult to find in the microstructure. This worth noting, because the DP600 steel exhibits a high amount of finely-dispersed martensitic inclusions with an average diameter of just 1 μm.

Nano-indentation. The micro-pillars were loaded with a U9820A Keysight Nano Indenter G200 (Keysight Technologies Inc., Santa Rosa, USA), utilizing truncate-cone-like indenter tips with sidewall angles of $60°$ and flat-punch diameters of 10 μm. “The deformation was applied with a constant strain rate of 1.6·10-3 s-1 up to a nominal displacement of 10 % of the original height (Figure 7.2). The dimensions of the micro-pillars before and after deformation were measured by SEM and the load and displacement values recorded by the indenter were used to calculate the engineering stress and strain curves.”Wenk (2014)

5.3 Subsequent reloading experiment

A uniaxial, strain-controlled tension experiment with subsequent reloading in the same direction was conducted by Simone Schuster[3] (Schuster, 2014) to investigate the microscopic evolution of the DP600 steel. The specimen had a shape similar to DIN specifications. The specimen had fine-ground surfaces and was strained by a Kammrath & Weiss 10 kN loading device with a mechanical strain-gauge. The loading device was installed within a X-ray diffraction device (Cr-K α, 211-peak, Institute for Applied Materials, Karlsruhe Institute for Technology). The detection signal was evaluated according to the $\sin^2\Psi$-method because the investigated DP600 exhibits only slightly anisotropic behavior. The slight anisotropy causes only acceptably small errors occur, using this method.

[3] Institute for Applied Materials – Materials Science (IAM-WK), DFG-GRK1483 Subproject A13

Subsequent measurements were conducted for the initial state and after elastic straining ($\varepsilon = 0.00072; 0.00143$) as well as after plastic straining ($\varepsilon = 0.01; 0.0225; 0.0355; 0.053; 0.076; 0.102; 0.153$):

- in-situ measurements of the ferrite stress (XRD, Figure 7.2a, Schuster (2014)),
- ex-situ measurements of the ferrite residual-stress in loading direction (XRD, Figure 7.2a, Schuster (2014)),
- integral orientation-distributions (XRD, not reproduced here, Schuster (2014)),
- local orientation-distributions (EBSD[4]), not reproduced here, Wenk (2014)) and
- the evolution of a small region of interest (EBSD, not reproduced here, Wenk (2014)).

The long-range-stress measurements below $\sim 1\%$ were not reliably interpretable due to pre-deformation and textural effects. Additionally, the chemical composition (Table 5.2) was measured with an ES 750-CA spark-emission spectrometer (Gesellschaft für Elektronik und Feinwerktechnik mbH, OBLF, Germany). DP600 is composed of ($\mathrm{wt.\%}$) Fe–$0.093\,\%$C–$0.3\,\%$Si–$1.5\,\%$Mn, the carbon content in martensite is estimated to be $C_{\mathrm{M}} = 0.66\,\%$. The DF140T steel from Chen et al. (2014) has Fe–$0.15\,\%$C–$0.3\,\%$Si–$1.5\,\%$Mn, the estimated carbon content in martensite is found to be $C_{\mathrm{M}} = 0.33\,\%$.

[4] all EBSD measurements by Moritz Wenk, Institute for Applied Materials – Materials- and Biomechanics (IAM-WBM), DFG-GRK1483 Subproject A11

Chapter 6

Mechanism based mean-field material model

In this section, an original material-model for DP steels is derived in order to model incompatibility-induced long-range stresses that interact with both ferrite and martensite[1]. In essence, the model combines the fundamental works of Ashby (1970) and Brown and Stobbs (1971a) with the composite model of Mughrabi (1983).

In view of the quasi-isotropic behavior of the DP600 steel (Section 5.1), the proposed material model is assumed to be isotropic and makes use of the standard von Mises model described in Section 2.3. Note that only rate-independent strain-hardening is considered. Strain-rate hardening or strain-rate sensitivity is briefly discussed in Section 6.1. Strain softening may also be observed in metals, e.g., at higher temperatures (DeLo and Semiatin, 1999) or in nanocrystalline metals (Tang and Schoenung, 2008). Finally, instability in plastic flow is sometimes present in metals for certain strain-rate and temperature combinations, e.g., in form of the Portevin-Le Chatelier effect (e.g., Böhlke et al., 2009). A serrated flow-curve is characteristical for the Portevin-Le Chatelier effect if it occurs due to dislocations being temporarily ar-

[1] This section is reproduced from the submitted article: F. Rieger, T. Böhlke, Mechanism based mean-field modeling of the work-hardening behavior of dual-phase steels.

rested by fast diffusing dilute atoms (Kubin et al., 2002). Queiroz et al. (2012) observe serrated flow in a DP600 steel at temperatures between 155 °C and 250 °C and strain-rates of $10^{-4} - 10^{-2}\,\mathrm{s}^{-1}$. They conclude that DP steels exhibit similar dynamic strain-aging as (unalloyed or alloyed) low-carbon steels. In their investigated DP steel, dynamic strain-aging is caused by the locking of dislocations by carbon atoms in ferrite. The consideration of strain-rate dependent hardening or instabilities would be applicable for thermo-mechanical models, but is omitted, in the work at hand. For example, the dissipation heat from plastic deformation might lead to higher temperatures in DP steels during forming operations. In combination with the high strain-rates, the parameter window for the occurrence of the Portevin-Le Chatelier effect could be met during automotive forming operations.

6.1 Dislocation-density based modeling

Taylor yield-stress relation. The isotropic flow-stress σ^{f} of the model is based on the fundamental and widely accepted Taylor relation

$$\sigma^{\mathrm{f}} = \sigma^0 + MGb\alpha\sqrt{\rho}, \tag{6.1}$$

accompanied by constant flow-stress contributions that are gathered in σ^0. The constant stress $\sigma^0 = \sigma^0_{\mathrm{PF}} + \sigma^0_{\mathrm{SS}} + \sigma^0_{\mathrm{GS}}$ covers different effects: the Peierls friction-stress by σ^0_{PF}, solid-solution strengthening by σ^0_{SS}, and grain-size effects by σ^0_{GS} (see, e.g., Dunstan and Bushby (2014) for a critical discussion of the grain-size effect on initial yield-strength). Equation (6.1) includes the experimentally measurable average dislocation-density ρ as the sole internal variable. Here, $M \simeq 3$ is taken as Taylor factor for a random orientation-distribution. It relates the microscopic resolved shear-stress to the macroscopic uniaxial-stress via $\sigma = M\tau$. The isotropic shear-modulus is given by G, and b

is the Burgers vector's magnitude. With Equation (6.1), the material's strength is solely determined by dislocation-dislocation interactions, for which $\alpha \simeq 0.3$ is a dimensionless constant (Delince et al., 2007). Interactions are assumed to be orientation-independent or isotropic.

Kocks-Mecking dislocation-density evolution. The dislocation density evolves with ongoing plastic deformation. A common evolution law (Kocks and Mecking, 2003) is given by the relation

$$\frac{\mathrm{d}\rho}{\mathrm{d}\varepsilon^{\mathrm{p}}} = M\left(k^{\mathrm{f}}\sqrt{\rho} - \hat{k}^{\mathrm{a}}\rho\right). \tag{6.2}$$

The dislocation storage term $k^{\mathrm{f}}\sqrt{\rho}$, depends solely on the mean-free traveling path of dislocations prior to their immobilization, e.g., by dislocation networks. For a random dislocation-arrangement, the average dislocation-spacing is proportional to $1/\sqrt{\rho}$ and k^{f} is a storage constant that is related to the number of cut dislocations before immobilization finally occurs. The term $k^{\mathrm{f}}\sqrt{\rho}$, therefore, quantifies statistically-stored dislocations in the sense of Ashby (1970).

Dynamic recovery is modeled to be proportional to the current dislocation density by $\hat{k}^{\mathrm{a}}\rho$. It is determined by, e.g., annihilation through cross-slip of screw dislocations or climb of edge dislocations (Estrin, 1996). The dominant mechanisms in this process are temperature and strain-rate dependent, as is the recovery parameter $\hat{k}^{\mathrm{a}}$. However, the common assumption that Equation (6.2) is temperature and strain-rate *independent* is adopted for the small strain-rates at room temperature in this work. The relation inEquation (6.2) is originally intended for face-centered cubic metals, but has widely been employed for body-centered cubic metals in the past. On this account, it is remarked that the two lattice structures behave similarly if the temperature is high enough for body-centered metals to be governed by dislocation-dislocation interactions (Kocks and Mecking, 2003).

Strain-rate sensitivity. Regarding the measured strain-rate dependence of the DP600 steel (Figure 5.1), a modification to the recovery coefficient k^{a} could be found in the works of Estrin (1996); Kocks and Mecking (2003):

$$k^{\mathrm{a}} = k_0^{\mathrm{a}} \left(\frac{\dot{\varepsilon}_0}{\dot{\varepsilon}^{\mathrm{p}}} \right)^n . \tag{6.3}$$

Equation (6.3) is based on an Arrhenius law of thermal activation with a logarithmically stress-dependent activation energy. This allows for identifying the exponent $n = kT/A$ as a function of temperature T. The material constant A depends on the relative stacking-fault energy and k is the Boltzmann constant. Another variant of incorporating strain-rate and temperature dependence is to scale a reference work-hardening $\sigma_{\dot{\varepsilon}_0}$ by influence functions. A variety of possible influence functions for this modeling strategy are discussed in Sung et al. (2010) for DP steels. Additionally, experimental data for DP600 steels from Beynon et al. (2005) for high strain rates of $\dot{\varepsilon} = 10^2\,\mathrm{s}^{-1}$ is considered on this account, here. It is found that the simple phenomenological strain-rate dependence laws (the power-law model and the Johnson-Cook model, as cited in Sung et al. (2010)) cannot reproduce the complete experimental data. However, the more complex model of Wagoner (1981), the expression

$$\sigma = \sigma_{\dot{\varepsilon}_0} \left(\frac{\dot{\varepsilon}^{\mathrm{p}}}{\dot{\varepsilon}_0} \right)^{m(\dot{\varepsilon} \cdot \dot{\varepsilon}_0)^n} , \tag{6.4}$$

is able to fit the experimental data over six orders of magnitude (Figure 6.1). To get a first estimation of the strain-dependent DP steel behavior, one could use the relation (Sung et al., 2010)

$$\sigma^{\mathrm{f}} = \left(\sigma^0 + MGb\alpha\sqrt{\rho} \right) \left(\frac{\dot{\varepsilon}^{\mathrm{p}}}{\dot{\varepsilon}_0} \right)^{m(\dot{\varepsilon} \cdot \dot{\varepsilon}_0)^n} . \tag{6.5}$$

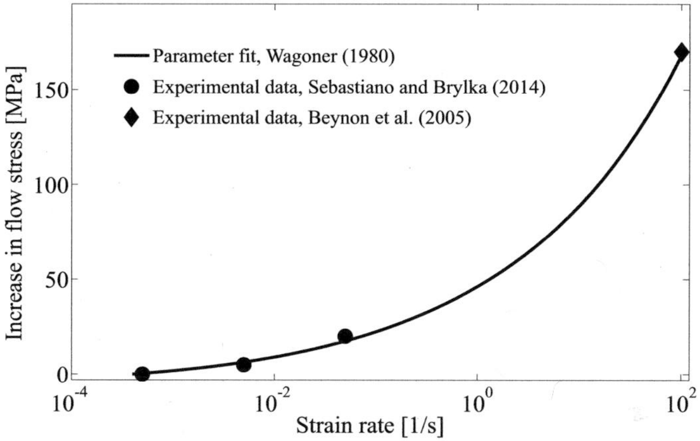

Figure 6.1: Fit of the Wagoner (1981) model to strain-rate data for DP600 steels (experimental data as of Section 5.1 and Beynon et al., 2005). The fitted parameters are $\dot{\varepsilon}_0 = 5 \cdot 10^{-4}\,\mathrm{s}^{-1}$, $m = 0.048$ and $n = 0.3$. The power-law model and Johnson-Cook model (as cited in Sung et al. (2010)) were not appropriate to fit the data over six decades.

Sung et al. (2010) also give different models for the temperature scaling. However, as there is no experimental data available for the DP600 steel, this effect is not considered, here. Han et al. (2005) find no consistent effect of strain rate on the long-range stresses for reverse-loading (Bauschinger) tests in DP steels. Within the tested range of strain-rates ($10^{-4} - 10^{-2}$), uncertainty could not be ruled out. Their measured values appear to scatter randomly within $\sim 20\,\%$.

Influence of microstructure. Equation (6.2) assumes an average grain-size much larger than the average dislocation-spacing. For such polycrystals, grain-boundaries do not significantly contribute to dislocation immobilization. With grain sizes below $\sim 30\,\mu\mathrm{m}$, however, there is a distinct effect of grain size on the constant stress σ^0, and the work-hardening rate. Therefore, it appears to be necessary to introduce a third term to Equation (6.2), as previously suggested by Estrin (1996); Kocks and Mecking (2003):

$$\frac{\mathrm{d}\rho}{\mathrm{d}\varepsilon^{\mathrm{p}}} = M\left(\beta\left(\rho, \mathrm{MS}\right) + k^{\mathrm{f}}\sqrt{\rho} - \hat{k}^{\mathrm{a}}\rho\right). \tag{6.6}$$

The natural dependencies of β on the current internal state (indicated by ρ) and on the microstructure (indicated by MS) are often neglected in the literature. However, both dependencies are present in DP steels and, thus, the proposed β-term incorporates different storage processes for geometrically-necessary dislocations. The original formulation of $\beta = (bD)^{-1}$ is based in the assumption that grain-boundaries directly limit the dislocation mean-free path. No additional effects, like inclusions, are considered in the original formulation, however.

From Equation (6.6), three basic questions remain open and will be discussed in the following. Firstly, the dislocation annihilation term $-\hat{k}^{\mathrm{a}}\rho$ is an ad-hoc assumption and the resulting linear reduction of the work-hardening rate is only an approximation of the experimentally observed behavior; albeit over a quite large stress regime. Secondly, to the best knowledge of the author, the dependencies of β on plastic state and microstructure have not been satisfactorily modeled in the literature, yet. Especially the microstructure influences the work-hardening behavior of DP steels substantially (e.g., Park et al., 2014; He et al., 1984). A physically motivated (averaged) description is, thus, necessary for this work. Thirdly, the interaction of ferrite and martensite through long-range stresses and the resulting load transfer from ferrite to martensite, has not been modeled in detail with average dislocation-density models, yet.

Modeling in this work, deliberately, assumes spherical grains, exclusively. In addition, all microstructural parameters, e.g., the grain size, are used in an average sense. There are findings, however, that at least for grain size, this is a major concession to simplicity (see, Böhlke et al. (2014) or the investigation in Chapter 3).

6.2 Exponential dynamic recovery

The first modification to Equation (6.6) presented in the following, is an alternative dynamic-recovery formulation. Following the work of Bouaziz (2012), an exponential decrease of the work-hardening is used

$$\frac{\mathrm{d}\rho}{\mathrm{d}\varepsilon^{\mathrm{p}}} = M\left(\beta\left(\rho, \mathrm{MS}\right) + k^{\mathrm{f}}\sqrt{\rho}\right) \cdot \exp\left(-k^{a}\sqrt{\rho}\right). \tag{6.7}$$

The constant k^{a} is interpreted as the capture distance for dynamic recovery. At the point of initial plastic deformation, the new formulation recovers the original Kocks-Mecking evolution law (Equation (6.2)). For subsequent plastic deformation, however, it retains a higher work-hardening rate over the Kocks-Mecking model. The reason for this behavior is the exponential decay of the new model which approaches zero work-hardening asymptotically. The Kocks-Mecking work-hardening decay is linear and the Kocks-Mecking model actually shows absent work-hardening, starting from a certain high plastic-deformation onwards. Parameter fits to the micro-pillar compression tests also show better accuracy for the new formulation in Equation (6.2), supporting that an exponential decay is closer to experiments for single-phase steels (Kocks and Mecking, 2003; Bouaziz, 2012). Essentially, Equation (6.7) allows for a better representation of stage IV work-hardening at high strains without introducing any additional parameters. In general, dynamic recovery governs the high-strain behavior. The formulation of Equation (6.7) has been originally developed for high strain behavior. It is commonly accepted, that the high-strain behavior of DP steels is similar to that of single-phase materials (Korzekwa et al., 1984; Ramazani et al., 2013). As it is common in the literature, the different mechanisms of dynamic recovery are not treated individually, here.

6.3 Isotropic hardening of plastically non-homogeneous materials

6.3.1 Theoretical basis for abstracted geometries

In a second step, the theoretical findings of Ashby (1970) are used to acquire a physically motivated dislocation-density storage. With the findings of Ashby (1970) for the averaged dislocation density storage, it is possible to incorporate the microstructure of the respective DP steel. Ashby (1970) considers a ductile matrix interspersed with strongly bonded, small and rigid particles. He identifies three fundamental mechanisms of dislocation storage in plastically non-homogeneous materials when the incompatibility is fully plastically relaxed:

Martensitic particles. The first mechanism is the formation of geometrically necessary dislocations to preserve the compatibility of non-deforming particles with a plastically-deformed surrounding matrix. From theoretical and geometrical considerations, one finds the geometrically-necessary dislocation density ρ^G to scale with plastic shear by the well known relation of Ashby (1970),

$$\mathrm{d}\rho^{\mathrm{G}}_{\mathrm{MP}} \propto \frac{8c_{\mathrm{M}}}{bD_{\mathrm{M}}}\mathrm{d}\gamma. \tag{6.8}$$

For DP steels, harder particles are martensite particles (MP) with the particle diameter D_{M} and the martensite volume-fraction c_{M}. Only the proportionality is adopted, here, since this estimation is only valid for small plastic deformations.

Ferrite-ferrite grain boundaries. The second type of dislocation production originates from the incompatibility at single-phase grain boundaries. It is caused by different crystallographic orientations between neighboring grains. Single-phase grain boundaries can be identified as ferrite-ferrite (FF) grain boundaries for DP steels. The geometrically-necessary dislocation storage from this mechanism is found to scale proportionally with plastic strain by (Ashby, 1970)

$$\mathrm{d}\rho_{\mathrm{FF}}^{\mathrm{G}} \propto \frac{1}{4bD_{\mathrm{F}}}\mathrm{d}\gamma. \tag{6.9}$$

Compared to Equation (6.8), this ferrite-ferrite contribution is much smaller in DP steels. Reasons are the much smaller geometrical factor (the value is $1/4$) and that fact that ferrite grains are often larger than martensite grains in dual-phase steels.

Ferrite-martensite phase boundaries. The third fundamental contribution to the β-term originates from dislocations that get trapped inside matrix-grains which are completely covered by a significantly harder phase. Dislocations then pile-up against the surrounding hard phase, facilitating long-range stresses in both phases (as is discussed later). The storage rate for this process is found to scale with plastic strain by (Ashby, 1970)

$$\mathrm{d}\rho_{\mathrm{FM}}^{\mathrm{G}} \propto \frac{4}{bD_{\mathrm{F}}}\mathrm{d}\gamma. \tag{6.10}$$

This mechanism can be related to the microstructure of DP steels by the analogy of a ferrite grain that is completely covered with martensite, i.e., a ferrite grain that has exclusively ferrite-martensite (FM) phase-boundaries. The quantitative difference between Equation (6.9) and Equation (6.10) matches experimental observations that ferrite-martensite phase-boundaries exhibit much larger dislocation pile-ups compared to ferrite-ferrite boundaries (Calcagnotto et al., 2010a; Tsipouridis et al., 2011; Kim et al., 2012). Tsipouridis et al. (2011)

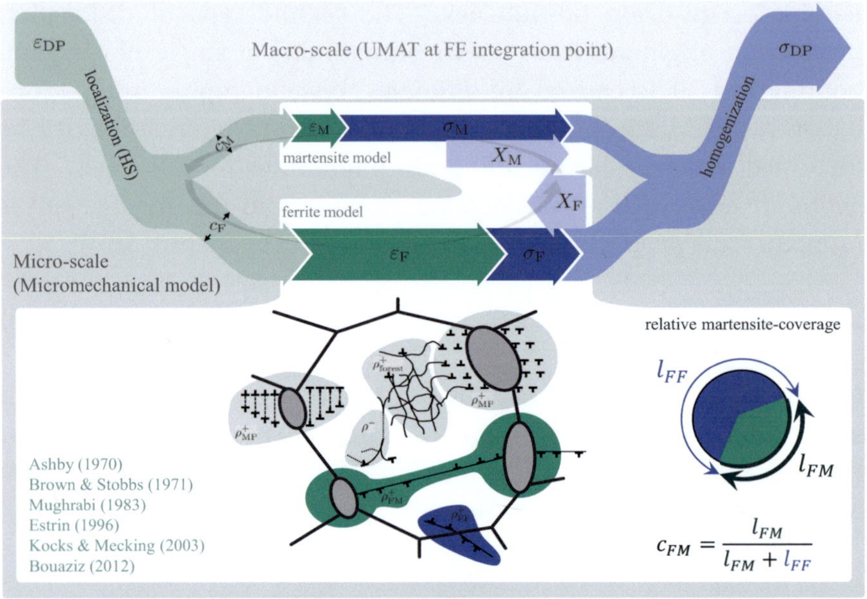

Figure 6.2: Schematic of the proposed model for DP steel. Macroscopic load is localized by a Hashin-Shtrikman based mean-field model. The classic storage term ρ^{+}_{forest} and the dynamic recovery ρ^{-} are taken from single-phase materials. Note the different additional dislocation-storage contributions, resulting from martensite particles ρ_{MP}, pile-ups at ferrite-martensite phase boundaries ρ_{FM}, and incompatibility at ferrite-ferrite grain boundaries ρ_{FF}. The former two dislocation arrangements (MP/FM) produce long-range stresses that act against the external load in ferrite $\boldsymbol{X}^{-}_{\text{F}}$ and with the external load in martensite $\boldsymbol{X}^{+}_{\text{M}}$.

explain the increased ferrite-martensite interface resistance with the change in crystal structure and the already high dislocation-density in martensite after the production process. Both factors impede dislocations from traversing the ferrite-martensite phase-boundary.

Equation (6.8) has been widely used for the description of dispersion-strengthening and DP steels, see, e.g., Brown and Stobbs (1971a), Balliger and Gladman (1981), Lanzillotto and Pickering (1982) or Bouaziz (2001). To the best knowledge of the author, however, all relations in

Equations (Equation (6.8), Equation (6.9), Equation (6.10)) have never been used jointly to model the work hardening of DP steel so far. This is surprising because, in principle, all three exemplary storage cases are equally present in DP steel. A material model that incorporates all three contributions is derived in the following. It is schematically depicted in Figure 6.2.

6.3.2 Model for dual-phase steels

For real microstructures, DP steels exhibit a less abstract microstructure. Parts of the ferrite grain boundary are of the ferrite-ferrite type and only a fraction is covered by martensite grains. For modeling the different grain-interfaces, the relative martensite-coverage ($c_{\mathrm{FM}} = l_{\mathrm{FM}} / \left(l_{\mathrm{FF}} + l_{\mathrm{FM}}\right)$) is used in the same form as it is proposed in Table 4.3. Contributions of the different grain-boundary types, ferrite-ferrite and ferrite-martensite, scale with their respective grain-boundary lengths. The relative length of ferrite-ferrite grain-boundaries is given by $c_{\mathrm{FF}} = 1 - c_{\mathrm{FM}}$. Using the three abstract and exemplary relations (Equation (6.8)-Equation (6.10)), the β-term is assembled by

$$\beta\left(\rho_{\mathrm{F}}, MS\right) = \underbrace{\frac{8}{\pi b D_{\mathrm{M}}} c_{\mathrm{M}}}_{\beta_{\mathrm{MP}}} + \underbrace{\frac{4}{b D_{\mathrm{F}}} c_{\mathrm{FM}}}_{\beta_{\mathrm{FM}}} + \underbrace{\frac{1}{4 b D_{\mathrm{F}}} \left(1 - c_{\mathrm{FM}}\right)}_{\beta_{\mathrm{FF}}} . \tag{6.11}$$

It is used for the dislocation-density evolution

$$\frac{\mathrm{d}\rho}{\mathrm{d}\varepsilon^{\mathrm{p}}} = M \left(\beta\left(\rho_{\mathrm{F}}, MS\right) + k^{\mathrm{f}} \sqrt{\rho_{\mathrm{F}}}\right) \exp\left(-k^{a} \sqrt{\rho_{\mathrm{F}}}\right) . \tag{6.12}$$

Although the contribution of $\left(4 b D_{\mathrm{F}}\right)^{-1}$ to the overall dislocation production in ferrite ρ_{F} is often comparably small (Jiang et al., 1992), it appears to be natural and without drawbacks to incorporate it, here. Inverse scaling with $1/bD$ is common for a variety of formulations in literature (e.g., Ashby, 1970; Brown and Stobbs, 1971a; Jiang et al., 1995; Bouaziz, 2001; Delince et al., 2007). The formulation of Equation (6.11) assumes a random dislocation distribution for the annihilation.

The work of Ashby (1970) assumes very thin plates of hard material in Equation (6.10). In DP steels, though, martensite covers ferrite grain-boundaries in the form of many small, spherical grains (Figure 7.5). Nevertheless, the principal mechanism of the dislocation pile-up against the harder phase is analogous to the theoretical hypothesis: the surrounding hard phase (at least partially) hinders the deformation of the enclosed ferrite grain. The effect of martensitic particles in Equation (6.11) is, thus, twofold: due to their incompatibility with the surrounding matrix (see Equation (6.8)), martensitic particles introduce geometrically-necessary dislocations themselves. Secondly, the movement of statistical dislocations inside the ferrite-grain dislocation is hindered. This constraint leads to additional dislocation storage when dislocations pile-up against the covering martensite phase (see Equation (6.10)).

Regarding Equation (6.11), there is a factor of two when compared to the derived relations in Ashby (1970) and Brown and Stobbs (1971a). The exact numerical factors for each contribution depend on geometry and the dislocation arrangement itself, which are incorporated in a simplified way, here. Essentially, the theoretical derivations capture the mechanisms quantitatively for the abstract cases and it is assumed that the different terms in Equation (6.11) quantify the *relative* contribution to the overall behavior. As both mentioned works postulate assumptions for this factor, it could be considered as a fitting parameter. In this work, however, the more commonly adopted version of Ashby (1970) is used, instead. The chosen formulation allows for better overall fits and, especially, better fits to the experimentally determined microscopical and macroscopical data for the DP600 steel (Section 7.1).

As it is common in the literature, plain additivity of different dislocation-storage contributions (e.g., Estrin (1996), Jiang et al. (1992)) is assumed in Equation (6.34). An interaction of the different contributions is experimentally observed (Brown and Stobbs, 1971a), but an additive relation is clearly a simplification of the actual spatial dislocation interactions. A certain fraction of geometrically-necessary dislocations contributes directly to the Taylor-type flow stress in the three-dimensional case. It is, however, difficult to quantify this fraction, exactly. The main issue is to develop a suitable spatial averaging procedure and a representation of the complex dislocation arrangement; if the arrangement would be known in detail (Mughrabi, 2001).

Brown and Stobbs (1971a) observed interactions between, what they call, the „forest of GNDs" and statistical dislocations from the grain interior. Additionally, secondary dislocations interact with each other within the localized dislocation-entanglements around particles. Additivity in Equation (6.34) is equivalent to a full interaction between GNDs from the β-term and statistically-stored dislocations from the grain interior ($k^{\mathrm{f}}\sqrt{\rho}$). Even an averaged spatial resolution, e.g., one hard and one soft ferrite-region as done in Pipard et al. (2009), would increase the model complexity, significantly.

Regarding the real dislocation arrangements, dislocations near particles follow a certain trend (Brown and Stobbs, 1971a): Small particles of diameters $\ll 1\,\mu\mathrm{m}$ exhibit primary slip for small deformations of more than 5%. For higher deformations or larger particles of diameters $> 1\,\mu\mathrm{m}$, the primary slip mechanism is unfavorable compared to complex secondary slip entanglements. Thus, complex arrangements dominate the behavior for DP steels as they exhibit large particles and undergo high deformations. The irregular structure of geometrically-necessary dislocations around martensite particles resembles the statistically homogeneous distributions of dislocations. Such statistically homogenous interactions are described by plain additivity.

6.4 Kinematic hardening of plastically non-homogeneous materials

6.4.1 Long-range stress of dislocation arrangements

Plastic relaxation of the aforementioned particle incompatibility is mainly accomplished by formation of primary or secondary prismatic-loops. The plain presence of secondary loops already impedes the movement of primary dislocations, resulting in a local Taylor-type work-hardening. Secondary dislocation loops impede further plastic relaxation because the necessary stress for primary dislocations to cut through entanglements increases. This facilitates local elastic stresses and long-range stresses. The primary dislocations, at least in part, exhibit a dipole structure centered around particles, as it is described in Brown and Stobbs (1971a).

Geometrically-necessary dislocation pile-ups with a dipole structure can induce two different effects: if fully relaxed, they induce crystallographic curvature without long-range stresses; if fully constrained, they induce long-range stresses without curvature (Mughrabi, 2006). Like most materials, DP steels exhibit an intermediate behavior. Curvature can be observed together with long-range stresses (Dillien et al., 2010; Nesterova et al., 2015). This corresponds to a partly relaxation of the long-range stresses. The herein proposed model takes this effect into consideration by introducing the factor f_{LRS}. This factor gives the fraction of dislocations around a particle that actually constitute long-range stresses (unrelaxed strain) and is derived in what follows. Note that f_{LRS} is not included in the isotropic forest-hardening term. This essentially reflects the duality of geometrically-necessary dislocation pile-ups. They might introduce long-range stresses which partly relax but at the same time they also hinder other dislocations in a grain from passing the arrangement.

For the quantification of the long-range stresses, the basic model of Brown and Stobbs (1971a) is extended to be applicable for DP steels. Brown and Stobbs (1971a) derive a quantitative estimation for the unrelaxed strain $\tilde{\varepsilon}^{\mathrm{p}}$ that causes long-range stresses. The derivation is briefly summarized here, for the sake of completeness.

The number of GND loops n is estimated with the extended Ashby (1970)-formula including the unrelaxed shear-strain $\tilde{\gamma}^{\mathrm{p}}$ by

$$n = 2\left(\gamma^{\mathrm{p}} - \tilde{\gamma}^{\mathrm{p}}\right)\frac{D_{\mathrm{M}}}{b}. \tag{6.13}$$

In this formula, the term $\gamma^{\mathrm{p}} - \tilde{\gamma}^{\mathrm{p}}$ denotes strain that has "relaxed" into GNDs (b is the Burgers vector). Ashby's original formulation is valid for the estimation of GNDs at small strains. The formulation assumes full plastic relaxation with $\tilde{\gamma}^{\mathrm{p}} = 0$. Long-range stresses, however, stem from unrelaxed incompatibilities. From the assumption that each prismatic loop possesses a diameter close to the reference-particle size follows

$$\rho_{\mathrm{local}} = 4n\frac{D_{\mathrm{M}}}{b}. \tag{6.14}$$

A crucial step of their derivation lies in utilizing the shear-stress equality at the particle interface. The particle is subjected to a uniform shear strain of $-\tilde{\gamma}^{\mathrm{p}}$, if the surrounding matrix is sheared by $\tilde{\gamma}^{\mathrm{p}}$ (Ashby, 1970; Brown and Stobbs, 1971a). Thus, the stress inside a particle,

$$\tau_{\mathrm{MP}} = G\tilde{\gamma}^{\mathrm{p}}, \tag{6.15}$$

is constant throughout the particle and, consequently, it is the same as on the interface. On the interface, however, the stress from unrelaxed dislocations is equal to the local yield-stress of ferrite, i.e., this yields

$$\tau_{\mathrm{local}} = Gb\alpha\sqrt{\rho_{\mathrm{local}}}. \tag{6.16}$$

If the stress would be higher than the local yield-stress, the material would relax plastically. From this, one obtains a description of the unrelaxed plastic strain from Equation (6.13)-Equation (6.16) that reads

$$G\tilde{\gamma}^{\mathrm{p}} = Gb\alpha\sqrt{8\left(\gamma^{\mathrm{p}} - \tilde{\gamma}^{\mathrm{p}}\right) D_{\mathrm{M}}^{2} / \left(bV_{\mathrm{p}}\right)}. \tag{6.17}$$

Assuming a plate-like shape (with main-diameter D_{p}) the volume reads

$$V_{\mathrm{p}} = \tfrac{\pi}{8} D_{\mathrm{p}}^{2} D_{\mathrm{M}}. \tag{6.18}$$

The link of a dislocation at the edge of the zone produces slip. Leaving the zone is considered as a behavior similar to a Frank-Read source with an activation stress of $\sigma_{\mathrm{act}} = 2Gb/D_{\mathrm{M}}$. Due to the high local elastic-stresses in the vicinity of particles, Frank-Read sources can be operative despite the small impeding distance D_{M}. Unrelaxed strains around particles produce a uniform long-range stress with an average value of $G\tilde{\gamma}^{\mathrm{p}}$ at the interface and within the particles (Ashby (1970), Brown and Stobbs (1971b)). The equality of the volume-averaged, locally elastic interface-stress σ_{int}, together with the critical stress for the operation of the Frank-Read source, is taken as a condition to compute the size of the plastic zone D_{p},

$$G\tilde{\gamma}^{\mathrm{p}} \tfrac{D_{\mathrm{M}}^{3}}{D_{\mathrm{p}}^{3}} = \frac{2Gb}{D_{\mathrm{M}}}. \tag{6.19}$$

Because most of the plastic strain is relayed, $\gamma^{\mathrm{p}} \gg \tilde{\gamma}^{\mathrm{p}}$, the size of the plastic zone in Equation (6.17) follows from Equation (6.17)-Equation (6.19), i.e.,

$$D_{\mathrm{p}} = \left(\frac{\alpha^2 \gamma^{\mathrm{p}} D_{\mathrm{M}}}{16\pi b}\right)^{\frac{1}{8}} 2D_{\mathrm{M}}. \tag{6.20}$$

With the size of the plastic zone D_{p}, the unrelaxed strain can be estimated by the expression

$$\tilde{\gamma}^{\mathrm{p}} = \left(\frac{16\pi b}{\alpha^2 \gamma^{\mathrm{p}} D_{\mathrm{M}}}\right)^{\frac{1}{8}} \alpha \sqrt{\frac{16 b \gamma^{\mathrm{p}}}{\pi D_{\mathrm{M}}}}. \tag{6.21}$$

Finally, a physically motivated estimation of the long-range stress magnitude is given by $G\tilde{\gamma}^{\mathrm{p}}$which leads to the relation

$$\tau^{\mathrm{LRS}} = G\tilde{\gamma}^{\mathrm{p}} \simeq \left(\frac{16\pi b}{\alpha^2 \varepsilon^{\mathrm{p}} D_{\mathrm{M}}}\right)^{\frac{1}{8}} G\alpha \sqrt{\frac{16 b \varepsilon^{\mathrm{p}}}{\pi D_{\mathrm{M}}}}. \tag{6.22}$$

The stress magnitude τ^{LRS} is not the average long-range stress in ferrite but the uniform stress inside the particle and, thus, at the particle interface. Note that the approximation for the volume average from Brown and Stobbs (1971b), $\tau_{\mathrm{F}}^{\mathrm{LRS}} \simeq c_{\mathrm{M}} G\tilde{\gamma}^{\mathrm{p}}$, is not adopted, here. A reasoning for this is given in Section 6.4.2.

The two different formulations, Equation (6.34) and Equation (6.22), are combined to apply the model to DP steel. Therefore, Equation (6.22) is rewritten for the case of uniaxial tension in terms of dislocation densities,

$$\sigma_{\mathrm{M}}^{\mathrm{LRS}} = MGb\alpha\sqrt{\rho_{\mathrm{LRS}}}. \tag{6.23}$$

By this, the evolution equation of the dislocation-density from Brown and Stobbs (1971b) is then given by

$$\frac{\mathrm{d}\rho_{\mathrm{LRS}}}{\mathrm{d}\varepsilon_{\mathrm{F}}^{\mathrm{p}}} = M \left[\frac{8}{\pi b D_{\mathrm{M}}}\right] f_{\mathrm{LRS}}, \tag{6.24}$$

with the factor

$$f_{\mathrm{LRS}} = \frac{3}{4}\left(\frac{16\pi b}{\alpha^2 D_{\mathrm{M}} \varepsilon_{\mathrm{F}}^{\mathrm{p}}}\right)^{\frac{1}{4}}. \tag{6.25}$$

The contributions of martensite particles to ρ_{F} and ρ_{LRS} are identical to the original work. Equation (6.23) has the same form as the Taylor-relation in Equation (6.1), although it has been derived for long-range stresses and not for forest hardening. This is due to the assumption that the local stress at the interface is limited by the current local strength. And the local strength, in turn, is defined by the Taylor-relation for this approach.

6.4.2 Extension to dual-phase steels

The model of Brown and Stobbs (1971a) has three shortcomings for modeling DP steels. Firstly, the factor c_{M} is an approximation for the small particle volume-fractions of the material investigated in their work. Additionally, their precipitate-hardenend material exhibits no grain-boundary coverage as it is described in Section 6.3.1. In other words, DP steels exhibit significantly higher martensite volume-fractions up to $60\,\%$ and possess a more complex (less theoretical) microstructure. Secondly, they appropriately assume undeformable particles, while in DP steels, martensite deforms at higher macroscopic strains (Mazinani and Poole, 2007a). And thirdly, Brown and Stobbs (1971a) do not include matrix-particle interactions because they assume rigid particles in their model.

Starting with the single-particle description from Equation (6.22), the composite model of Mughrabi (1983, 2001) allows for a more appropriate load-transfer model in the context of DP steel. The Mughrabi (1983) model assumes gliding dislocations that traverse through the grain interior. After doing so, the traversing dislocations create pile-ups at the interfaces between the hard and soft phases. For the simplified one-dimensional model of Mughrabi (1983) that is used here, the effective burgers vectors are opposite on adjacent sides of the hard phase. Similarly to Ashby (1970), additional dislocations must, therefore, be created at the interface to maintain compatibility between the two plastically inhomogeneous phases.

The dislocation behavior is described analogously in the Ashby (1970) and Mughrabi (1983) models: dislocations from the grain-interior pile-up against the soft-hard interface. As a direct conclusion of this modeling approach, the ferrite-martensite phase-boundary effect (Equation (6.10)) is to be incorporated into the evolution of dislocations that constitute long-range stresses (Equation (6.24)). Again, plain summation of the contributions is assumed, and thus,

$$\frac{\mathrm{d}\rho_{\mathrm{LRS}}}{\mathrm{d}\varepsilon_{\mathrm{F}}^{\mathrm{p}}} = M\left[\frac{8}{\pi b D_{\mathrm{M}}} + \frac{4}{b D_{\mathrm{F}}} c_{\mathrm{FM}}\right] f_{\mathrm{LRS}}. \tag{6.26}$$

The martensite volume-fraction c_{M} is a global parameter, important for the volume-averaged long-range stresses in ferrite. However, it is not used in the balance of interfacial stresses, the basis of Equation (6.26). The stress balance of Equation (6.26) predicts local effects at ferrite-martensite phase-boundaries and the relative martensite-coverage c_{FM} is, thus, the averaged value of a local characteristic. As dislocations predominantly pile-up at ferrite-martensite interfaces, it appears to be natural to include c_{FM} in Equation (6.26). Additionally, there are long-range stresses associated with the pile-ups at ferrite-ferrite grain-boundaries (Sinclair et al., 2006; Delince et al., 2007) but their overall effect is small as quantified in Ashby (1970).

In the Mughrabi (1983) model, the dipolar nature of dislocation pile-ups constitutes long-range stresses that intensify the experienced load of the hard phase ("forward stress") and lessen the experienced load in the soft phase ("back stress"). As the macroscopic equilibrium must vanish for internal stresses, a quantitative relation between the two internal long-range stresses is given by (Mughrabi, 1983)

$$c_{\mathrm{F}}\sigma_{\mathrm{F}}^{\mathrm{LRS}} + c_{\mathrm{M}}\left(-\sigma_{\mathrm{M}}^{\mathrm{LRS}}\right) = 0. \tag{6.27}$$

Recall that the derivation of σ^{LRS} is, essentially, based on the equilibrium of the martensite-particle interface stress, which is identical to the constant stress inside martensite particles. Thus, $\sigma_{\mathrm{M}}^{\mathrm{LRS}}$ in Equation (6.27), denotes the long-range stress magnitude in martensite particles as calculated by Equation (6.22), i.e., $\sigma_{\mathrm{M}}^{\mathrm{LRS}} = \sigma^{\mathrm{LRS}}$. The negative sign in Equation (6.27) stems from the observation that long-range stresses reduce the bearable external load in the harder phase. Both effective scalar yield-stresses are then given by $\sigma_{\mathrm{F}}^{\mathrm{f,eff}} = \sigma_{\mathrm{F}}^{\mathrm{f}} + \sigma_{\mathrm{F}}^{\mathrm{LRS}}$ and $\sigma_{\mathrm{M}}^{\mathrm{f,eff}} = \sigma_{\mathrm{M}}^{\mathrm{f}} - \sigma_{\mathrm{M}}^{\mathrm{LRS}}$. However, a tensorial representation is used for the proposed model. For the average magnitude of long-range stress in the ferrite phase, the usage of Equation (6.27) together with $c_{\mathrm{F}} + c_{\mathrm{M}} = 1$ leads to the scalar relation

$$\sigma_{\mathrm{F}}^{\mathrm{LRS}} = \frac{c_{\mathrm{M}}}{c_{\mathrm{F}}}\sigma_{\mathrm{M}}^{\mathrm{LRS}} = \frac{c_{\mathrm{M}}}{1 - c_{\mathrm{M}}}\sigma_{\mathrm{M}}^{\mathrm{LRS}} = \frac{c_{\mathrm{M}}}{1 - c_{\mathrm{M}}}\sigma^{\mathrm{LRS}}. \tag{6.28}$$

Equation (6.28) extends the model to high martensite volume-fractions. For small volume-fractions, consequently, one finds $c_{\mathrm{M}}/c_{\mathrm{F}} \simeq c_{\mathrm{M}}$, and the approximation of Brown and Stobbs (1971a), $\sigma_{\mathrm{F}}^{\mathrm{LRS}} \simeq c_{\mathrm{M}}\sigma_{\mathrm{M}}^{\mathrm{LRS}}$ is recovered by the proposed relation.

The net effect of long-range stresses is a local load-redistribution from the softer ferrite to the harder martensite phase. According to the shear-lag analysis, deformation is accumulated in the ferrite matrix. Thus additional shear strains are created at the ferrite-martensite interfaces, redistributing load from ferrite to martensite. This load transfer causes an effective stress in martensite which is higher than the externally-applied stress and, in turn, relieves the ferrite parts in DP steels (Kim and Lee, 2000).

For the load-transfer modeling, the Armstrong-Frederick kinematic-hardening model is employed with the evolution equation (Frederick and Armstrong, 2007)

$$\dot{\boldsymbol{X}} = \tilde{c}\dot{\boldsymbol{\varepsilon}}^{\mathrm{p}} - \tilde{b}\dot{\varepsilon}^{\mathrm{p}}\boldsymbol{X}. \tag{6.29}$$

This relation makes use of the rate of the plastic-strain tensor $\dot{\boldsymbol{\varepsilon}}^{\mathrm{p}}$ and of the equivalent plastic strain $\dot{\varepsilon}^{\mathrm{p}}$; both are defined with regard to the isotropic von Mises model. Equation (6.29) is rewritten such that the kinematic-stress tensor $\boldsymbol{X}$ reads (e.g., Resende and Bouvier, 2013)

$$\mathrm{d}\boldsymbol{X} = c\left(X_{\mathrm{sat}}\left(\rho, \mathrm{MS}\right)\boldsymbol{N} - \boldsymbol{X}\right)\mathrm{d}\varepsilon^{\mathrm{p}}. \tag{6.30}$$

Equation (6.30) allows for a direct specification of the kinematic saturation stress, $X_{\mathrm{sat}} = \|\boldsymbol{X}_{\mathrm{sat}}\| = \sqrt{3\tilde{c}/(2\tilde{b})}$, which is the saturated magnitude for monotonic strains. A long-range stress definition in $\boldsymbol{N}$ corresponds to opposing the applied shear-stress on each slip system (Haddadi et al., 2006).

By using the previously deduced long-range stress magnitude (Equation (6.23), Equation (6.28)), an evolving saturation stress for uniaxial tension can easily be incorporated in the model, i.e.,

$$X_{\mathrm{sat}}\left(\rho, \mathrm{MS}\right) = \sqrt{\frac{2}{3}}\sigma_{\mathrm{F}}^{\mathrm{LRS}}. \tag{6.31}$$

This saturation stress function is a convenient choice for modeling $\boldsymbol{X}_{\mathrm{F}}$ (Section 6.4.2) due to its a scalar nature which suits the existing model of Brown and Stobbs (1971a).

The parameter $c = \tilde{b}$ defines the rate of adjustment (or the saturation rate) of the kinematic-stress tensor $\boldsymbol{X}$ to a modified loading direction, i.e., when the current kinematic-stress direction $\boldsymbol{X} / \|\boldsymbol{X}\|$ differs from the current effective-stress direction $\boldsymbol{N} = (\boldsymbol{\sigma}' - \boldsymbol{X}) / \|\boldsymbol{\sigma}' - \boldsymbol{X}\|$. The actual magnitude of the kinematic-stress tensor in Equation (6.30) is always smaller than X_{sat}. This behavior contradicts the definition of X_{sat} in Equation (6.31), since $\sigma_{\mathrm{F}}^{\mathrm{LRS}}$ is modeled to represent the *actual* current long-range stress magnitude in the material. Fortunately, with the high adjustment-rates in metals (e.g., $c \simeq 350$ for DP steel in Tarigopula et al. (2009)), the actual magnitude of the kinematic-stress tensor is very close to the saturation value. As the proposed model uses averaged values, it remains phenomenologically in its nature to a certain degree and the discrepancy of the kinematic stress magnitude, $\|\boldsymbol{X}\| - X_{\mathrm{sat}} < 0.1 X_{\mathrm{sat}}$, is within the overall uncertainty of this approach.

In real DP steels, a fraction of the fast-evolving incompatibility-dislocations becomes pinned at high strains due to secondary entanglements (Brown and Stobbs, 1971a). These dislocations are stored in slow evolving dislocation structures (Rauch et al., 2007, 2011). From the pinned (and slow evolving) dislocation structures, a fraction will recover during reverse- but not during orthogonal-straining in DP steels (Gardey et al., 2005). Compared to the fast reordering of dislocation pile-ups that constitute long-range-stresses, the rearrangement of entangled dislocation structures is delayed. A fraction of the dislocation structures created during forward loading is annihilated by reverse loading. This causes a transient hardening plateau during reverse loading in DP steels (and in certain interstitial-free ferrite steels).

For the primarily investigated DP600, data is only available for subsequent 90°-tensile loading. The exhibited hardening stagnation is very minor for this load case. This is in line with multiple observations from the literature for cross-loading in DP steels (Gardey et al., 2005; Tarigopula et al., 2008). Biaxial 90°-tensile loading is of mixed nature in between monotonic- and reverse-loading, and, thus, shows part of reverse-loading. The aforementioned absence of a hardening stagnation indicates that the overall effect for reverse-loading paths can be regarded as negligibly small, too. Therefore, the partial annihilation of entangled dislocation-arrangements is not included in the model. It is rather difficult to identify which fraction of which model-contribution is recovered during the rearrangement. With this in mind, an incorporation of this mechanism would introduce additional phenomenological parameters, i.e., at least the fraction of recovered dislocations (Rauch et al., 2007). Such experimental microstructural-data is not available for the investigated DP600 steel.

6.5 Saturation of incompatibility contributions

6.5.1 Kinematic hardening

The composite model of Mughrabi (1983) predicts a saturation of the interface-dislocation creation, i.e, a limit to the long-range stress. This saturation is also observed in DP steels. If sufficient load is transferred from the softer ferrite phase to the harder martensite phase, the latter one will inevitably yield. Consequently, the incompatibility of the two phases is certainly reduced because deformation becomes more uniform when both phases yield simultaneously. As a result, the creation of geometrically-necessary dislocations must decrease as well (Mughrabi, 1983). This effect has been described for DP steels by Sarosiek and Owen (1984): the initial behavior is dominated by the rapid creation of dislocations at the ferrite-martensite-interface. When dislocations have spread into the ferrite grain-interior and both phases deform almost uniformly, the work-hardening behavior of DP steels approaches the "regular" hardening of interstitial-free ferrite steels. The effect of plastically deforming martensite is implemented by a simple scaling law. Thus, with a scaling factor f_{FM}, the dislocation production rate from Equation (6.26) is extended for this effect and reads

$$\frac{\mathrm{d}\rho_{\mathrm{LRS}}}{\mathrm{d}\varepsilon_{\mathrm{F}}^{\mathrm{p}}} = M \left[\frac{8}{\pi b D_{\mathrm{M}}} + \frac{4}{b D_{\mathrm{F}}} c_{\mathrm{FM}} \right] f_{\mathrm{LRS}} f_{\mathrm{FM}}. \tag{6.32}$$

This scaling factor f_{FM} is, e.g., modeled by

$$f_{\mathrm{FM}} = 1 - \frac{\dot{\varepsilon}_{\mathrm{M}}^{\mathrm{p}}}{\dot{\varepsilon}_{\mathrm{F}}^{\mathrm{p}}}, \quad \text{with } \dot{\varepsilon}_{\mathrm{F}}^{\mathrm{p}} \neq 0. \tag{6.33}$$

Two properties of f_{FM} are essential. Firstly, it must not scale the dislocation production rate ($f_{\mathrm{FM}} = 1$) if martensite is elastic, i.e., $\dot{\varepsilon}_{\mathrm{M}}^{\mathrm{p}} = 0$. Secondly, long-range stresses must not increase ($f_{\mathrm{FM}} \simeq 0$) when both phases yield at equal rates, i.e., $\dot{\varepsilon}_{\mathrm{M}}^{\mathrm{p}} \simeq \dot{\varepsilon}_{\mathrm{M}}^{\mathrm{p}}$. Seyedrezai (2014) notes that "martensite particles still act as barriers to dislocation motion and therefore result in dislocation pile-ups despite their plastic deformation". Therefore, long-range stresses are still expected to rise slowly with increasing strain even after martensite has yielded. Exactly this behavior is predicted by the proposed model.

Another important physical constraint of the model can be deduced from the fact that the long-range stress magnitude must not exceed the (local) yield stress. If the long-range stress magnitude approaches the yield strength of ferrite, the production of long-range stresses should cease. Then, the material would simply yield to relax long-range stresses above the yield strength (see also Section 6.4.1). A possible factor f_{Y} to incorporate this physical bound analogously to f_{FM} would be, e.g., $f_{\mathrm{Y}} = \left(1 - \sigma^{\mathrm{LRS}}/\sigma^{\mathrm{f}}_{\mathrm{F}}\right)^{1/100}$. For the DP steels considered in this work, however, this physical bound is never violated, even without the usage of f_{Y}. This is probably the case, because it is already considered in the derivation of the long-range stress magnitude.

6.5.2 Isotropic hardening

There is also an effect of a more uniform phase-deformation on the dislocation storage for isotropic hardening in Equation (6.11). This effect is introduced using the same scaling factor f_{FM}, which then yields the extended relation

$$\beta\left(\rho, MS\right) = \frac{8}{\pi b D_{\mathrm{M}}} c_{\mathrm{M}} f_{\mathrm{FM}} + \frac{4}{b D_{\mathrm{F}}} c_{\mathrm{FM}} + \frac{1}{4 b D_{\mathrm{F}}} \left(1 - c_{\mathrm{FM}}\right). \tag{6.34}$$

The contribution is to be used in the dislocation evolution

$$\frac{\mathrm{d}\rho}{\mathrm{d}\varepsilon^{\mathrm{p}}} = M\left(\beta\left(\rho, MS\right) + k^{\mathrm{f}}\sqrt{\rho}\right)\exp\left(-k^{a}\sqrt{\rho}\right). \tag{6.35}$$

In Equation (6.34), the effect of ferrite-martensite interfaces ($4c_{\mathrm{FM}}/b/D_{\mathrm{F}}$) is not scaled down by f_{FM} after martensite yields. A physical motivation for doing so is given now. Dislocations from the grain interior still pile-up against the conglomerate of a martensite-covered grain-boundary and the work-hardened zone around it (the quote from Seyedrezai (2014) in Section 6.5.1 also applies, here). This effect does never stop completely. Even if both phases deform uniformly, the work-hardened zone is stronger than the completely unaffected ferrite grain-interior. Dislocations are then incorporated into the previously build entanglements and, by this mechanism, the dislocations are stored in the DP steel.

Constitutive model for the ferrite constituent

- Isotropic dislocation-density evolution (Ashby, 1970; Estrin, 1996; Bouaziz, 2012)

$$\frac{\mathrm{d}\rho_{\mathrm{F}}}{\mathrm{d}\varepsilon_{\mathrm{F}}^{\mathrm{p}}} = M \left[\underbrace{\frac{8}{\pi b D_{\mathrm{M}}} c_{\mathrm{M}} f_{\mathrm{FM}}}_{\text{mart. particles}} + \underbrace{\frac{4}{b D_{\mathrm{F}}} c_{\mathrm{FM}}}_{\rho \text{ restriction}} + \underbrace{\frac{1}{4 b D_{\mathrm{F}}} (1 - c_{\mathrm{FM}})}_{\text{FF incompatibility}} + \underbrace{k_{\mathrm{F}}^{\mathrm{f}} \sqrt{\rho_{\mathrm{F}}}}_{\text{forest hard.}} \right] \underbrace{\exp\left(-k_{\mathrm{F}}^{\mathrm{a}} \sqrt{\rho_{\mathrm{F}}}\right)}_{\text{dynamic recovery}}$$

- Saturation factor with progressively uniform deformation of both phases

$$f_{\mathrm{FM}} \overset{\text{e.g.}}{=} 1 - \frac{\dot{\varepsilon}_{\mathrm{M}}^{\mathrm{p}}}{\dot{\varepsilon}_{\mathrm{F}}^{\mathrm{p}}}$$

- Isotropic plastic comparison yield strength of Taylor-type

$$\sigma_{\mathrm{F}}^{\mathrm{f}} = \sigma^{0} + M G b \alpha \sqrt{\rho_{\mathrm{F}}}$$

- Incompatibility dislocations that constitute long-range stresses (Brown and Stobbs, 1971a)

$$\frac{\mathrm{d}\rho_{\mathrm{LRS}}}{\mathrm{d}\varepsilon_{\mathrm{F}}^{\mathrm{p}}} = M \left[\frac{8}{\pi b D_{\mathrm{M}}} + \frac{4}{b D_{\mathrm{F}}} c_{\mathrm{FM}} \right] f_{\mathrm{LRS}} f_{\mathrm{FM}}$$

- Fraction of long-range stress inducing dislocations(Brown and Stobbs, 1971a)

$$f_{\mathrm{LRS}} = \frac{3}{4} \left(\frac{16 \pi b}{\alpha^{2} D_{\mathrm{M}} \varepsilon_{\mathrm{F}}^{\mathrm{p}}} \right)^{\frac{1}{4}}$$

- Long-range stress magnitude (Mughrabi, 1983)

$$X_{\mathrm{sat}} = \sqrt{\frac{2}{3}} \sigma_{\mathrm{F}}^{\mathrm{LRS}} = \sqrt{\frac{2}{3}} \frac{c_{\mathrm{M}}}{1 - c_{\mathrm{M}}} M G b \alpha \sqrt{\rho_{\mathrm{LRS}}}$$

- Kinematic stress tensor evolution, $\boldsymbol{X}(0) = \mathbf{0}$ (Frederick and Armstrong, 2007)

$$\dot{\boldsymbol{X}}_{\mathrm{F}} = c \left(X_{\mathrm{sat}} (\rho, \mathrm{MS}) \boldsymbol{N} - \boldsymbol{X}_{\mathrm{F}} \right) \dot{\varepsilon}^{\mathrm{p}}$$

- Parameters: $\sigma_{\mathrm{F}}^{0}, \rho_{\mathrm{F}}^{0}, k_{\mathrm{F}}^{\mathrm{f}}, k_{\mathrm{F}}^{\mathrm{a}}$ and c

Constitutive model for the martensite constituent

- Dislocation density evolution

$$\frac{\mathrm{d}\rho_{\mathrm{M}}}{\mathrm{d}\varepsilon_{\mathrm{M}}^{\mathrm{p}}} = M \left[k_{\mathrm{M}}^{\mathrm{f}} \sqrt{\rho_{\mathrm{M}}} \right] \underbrace{\exp\left(-k_{\mathrm{M}}^{\mathrm{a}} \sqrt{\rho_{\mathrm{M}}}\right)}_{\text{dynamic recovery}}$$

- Isotropic plastic comparison yield strength of Taylor-type

$$\sigma_{\mathrm{M}}^{\mathrm{f}} = \sigma^{0} + M G b \alpha \sqrt{\rho_{\mathrm{M}}}$$

- Kinematic long-range stress tensor

$$\boldsymbol{X}_{\mathrm{M}} = -\frac{1 - c_{\mathrm{M}}}{c_{\mathrm{M}}} \boldsymbol{X}_{\mathrm{F}}$$

- Parameters: $\sigma_{\mathrm{M}}^{0}, \rho_{\mathrm{M}}^{0}, k_{\mathrm{M}}^{\mathrm{f}}, k_{\mathrm{M}}^{\mathrm{a}}$

Mean-field model

- Simplified Hashin-Shtrikman model

$$\Delta \boldsymbol{\sigma}_{\alpha} = \mathbb{C}_{\alpha} \left[\Delta \bar{\varepsilon} - \lambda \left\langle \Delta \varepsilon_{\alpha}^{\mathrm{p}} \right\rangle + (\lambda - 1) \Delta \varepsilon_{\alpha}^{\mathrm{p}} \right]$$

- The only parameter employed in this model is λ, defining the strain-partitioning

Figure 6.3: Summary of the proposed material model

6.6 Martensite material model

Compared to the ferrite phase, there is often less information available about the martensitic phase in DP steels. Reasons for this are manifold, e.g., the small martensite grain-size makes it difficult to gather information from EBSD measurements or micro-pillar compression tests. Then, the austenite-martensite phase-transformation is constrained by the surrounding ferrite matrix. In conjunction with the high number of martensite variants, this leads to a high defect density in martensite inclusions. A high defect density or high lattice distortion, in turn, cause a low image quality in EBSD measurements. Actually, martensite is sometimes identified by its low EBSD image quality (as done in Chapter 4 and Choi et al., 2013; Ramazani et al., 2013). With all these difficulties in mind, the martensite material-model is, deliberately, simple. It incorporates a similar dislocation-density evolution as for ferrite (Bouaziz, 2012),

$$\frac{\mathrm{d}\rho_{\mathrm{M}}}{\mathrm{d}\varepsilon_{\mathrm{M}}^{\mathrm{p}}} = M \left[k_{\mathrm{M}}^{\mathrm{f}} \sqrt{\rho_{\mathrm{M}}}\right] \exp\left(-k_{\mathrm{M}}^{\mathrm{a}} \sqrt{\rho_{\mathrm{M}}}\right). \tag{6.36}$$

The isotropic yield-stress matches the Taylor formulation for ferrite,

$$\sigma_{\mathrm{M}}^{\mathrm{f}} = \sigma^{0} + MGb\alpha\sqrt{\rho_{\mathrm{M}}}. \tag{6.37}$$

Kinematic stress that was build-up in ferrite ($\boldsymbol{X}_{\mathrm{F}}$) constitutes a kinematic stress inside martensite. The kinematic stress in ferrite has an opposing tensorial orientation and possesses a different magnitude. From Equation (6.27), it follows that

$$\boldsymbol{X}_{\mathrm{M}} = -\frac{1-c_{\mathrm{M}}}{c_{\mathrm{M}}} \boldsymbol{X}_{\mathrm{F}}. \tag{6.38}$$

With the interaction model from Equation (6.38), the stress equilibrium of Equation (6.27) is always satisfied for the complete stress tensor. Macroscopically, the average kinematic-stress tensor always vanishes. This also reflects that long-range stresses are entirely build up in ferrite but affect martensite, too. The complete material model is summarized in Figure 6.3 and a schematic can be found in Figure 6.2.

6.7 Parameter identification

The eight parameters ($\sigma^0_{\mathrm{F}}, \rho^0_{\mathrm{F}}, k^{\mathrm{f}}_{\mathrm{F}}, k^{\mathrm{a}}_{\mathrm{F}}$ and $\sigma^0_{\mathrm{M}}, \rho^0_{\mathrm{M}}, k^{\mathrm{f}}_{\mathrm{M}}, k^{\mathrm{a}}_{\mathrm{M}}$) for the two constituent behaviors are considered to be the minimal amount. Note that the Kocks-Mecking-Estrin model would need the same amount of parameters for two different materials. Additionally, the adjustment-rate parameter c for kinematic hardening has to be defined. The corresponding saturation parameter X_{sat} is expressed by means of geometrically-necessary dislocation pile-ups (Figure 6.3). Thus, a phenomenological parameter common to Armstrong-Frederick models is effectively eliminated by a physically-motivated relation. The mean-field model, finally, necessitates a last parameter λ. As is discussed in more detail in Chapter 2, this parameter governs the strain partitioning. The remaining parameters ($D_{\mathrm{M}}, c_{\mathrm{M}}, D_{\mathrm{F}}, c_{\mathrm{FM}}$) can be directly calculated from the microstructure, e.g., from two-dimensional EBSD data, as done in this work. These parameters represent leading geometrical micro-features of dual-phase steels.

Parameter identification for the DP600 model is based on the experimental data of Chapter 5. The model interactions are complex and highly non-linear, however, all of the material parameters have a physical interpretation. Taking into account all experimental data for each DP steel, respectively, it is possible to fit the physically-motivated material parameters manually (see Table 6.2).

Williams et al. (2013) state that "properties of martensite or ferrite in bulk form are not representative of their behavior at the microscale". On the other hand, the usage of micro-pillar compression tests for the constituents´ stress-strain curves imposes several problems, too. There is relatively high scatter in the experimental data, even for single phase materials. For DP steels the scatter is even higher, e.g., due to impurities from the respective other phase, or a different phase spatially right below a micro-pillar. Moreover, there is a distinct size effect in micro-sized specimen (Kraft et al., 2010). Micro-pillar compression tests give the range in which constituent properties must fall but it is nontrivial to derive „micro-bulk" phase properties from them (Ghassemi-Armaki et al., 2014).

The work-hardening behavior from the micro-pillar compression curves is averaged for the constituents, here (Kim et al., 2010). The initial constituent yield-strength is manually adjusted with respect to occuring size effects. A relatively soft phase behavior is motivated with a size effect similar to the grain- and micro-pillar diameter effects commonly observed in literature. The deliberately simple $1/D$-like strength dependencies do not distinguish between grain- and micro-pillar diameter (Schreijäg, 2013; Dunstan and Bushby, 2014). For the DP600 steel, the effective initial ferrite-constituent yield-stress of $\sim 260\,\mathrm{MPa}$ fits an extrapolation of the micro-pillar data based on this assumption (Wenk, 2014). The size effect of micro-pillars, thus, becomes small above $\sim 10\,\mu\mathrm{m}$ (Kiener et al., 2006).

There is additional XRD data for the ferrite constituent behavior of the DP600 steel. Comparison of XRD data with the micro-pillar compression data (Figure 7.2a) supports that the ferrite-phase behavior in this setup (including all DP work-hardening effects as measured by XRD) is soft compared to the micro-pillar compression tests. The actual ferrite-phase behavior should be even weaker than in the XRD measurements, because the characteristic DP work-hardening effects due to incompatibility are not included in the single-phase behavior.

Given the comparably high martensite carbon-content in the DP600 steel (see below), the micro-pillar compression tests of martensite exhibit relatively low strength. Therefore the modeled martensite-behavior dissents from the experimental data, however, the following reasoning is given:

During the production, carbon concentrates in the austenite phase and the carbon content is very low in ferrite. The carbon solubility limit of $C_{\mathrm{F}} \simeq 0.015\,\%$ is a first estimation of the carbon content in ferrite (Movahed et al. (2009), Byun and Kim (1993)). A simple and widely adopted estimation of the martensite carbon content, C_{M} (not to be mistaken with the martensite volume fraction c_{M}) can be deduced from the rule of mixtures as done in several works in the literature, e.g., Delince et al. (2007), Movahed et al. (2009), Ramazani et al. (2013) and Byun and Kim (1993),

$$C_{\mathrm{M}} = \frac{C_{\mathrm{DP}} - (1 - c_{\mathrm{M}})\, C_{\mathrm{F}}}{c_{\mathrm{M}}}. \tag{6.39}$$

For the investigated DP600 steel, Equation (6.39) yields $C_{\mathrm{M}} = 0.8\,\%$, which is comparably high. Therefore, instead $C_{\mathrm{M}} = 0.6\,\%$ is used as a rather conservative value. Linear relations might deviate for carbon contents above this value and there is always the possibility of carbon precipitates for such high carbon contents. A value of exactly $C_{\mathrm{M}} = 0.6\,\%$ has been estimated for a DP600 steel in Bergström et al. (2010), too; Marteau et al. (2013) even estimate a carbon content as high as $C_{\mathrm{M}} = 0.8\,\%$ for a DP600 steel.

To identify the initial martensite yield-strength, several empirical estimations can be found in the literature (Table 6.1). Note that the estimations differ for low carbon contents, although, they match nicely for

Formula [MPa]	σ_0^{f} [GPa]	$\%C_{\mathrm{DP}}$	$\%C_{\mathrm{M}}$	%Si	%Mn	**Reference**
$541 + 2289\sqrt{C_{\mathrm{M}}}$	$2.31 - 2.58$	$0.1 - 0.2$	RoM	2	1.5	Chen and Cheng (1989)
$900 + 1600 \cdot C_{\mathrm{M}}$	$1.86 - 2.18$	*)	$0.2 - 0.6$	0.25	0.85	Krauss (1995)
$635 + 2687 \cdot C_{\mathrm{M}}$	$2.24 - 2.78$	0.11	RoM	0.78	1.6	Byun and Kim (1993)
$1250 + 1600 \cdot C_{\mathrm{M}}$	$2.21 - 2.53$	0.15	RoM	0.5	1.7	Delince et al. (2007)
$77 + 3065 \cdot C_{\mathrm{M}}$	$2.12 - 2.52$	0.078	RoM	0.23	1.4	Gutiérrez and Altuna (2008)
–	2.2	0.093	RoM	0.29	1.6	this work

Table 6.1: Empirical extrapolations for martensite initial yield-strength from literature for low-carbon DP steels. $C_{\mathrm{M}} = 0.6 - 0.8\,\%$ has been used to calculate the range of σ_0^{f}. The chemical compositions and estimated martensite carbon contents are given for comparison. RoM indicates that the rule of mixtures has been used for the C_{M} calculation; *) indicates fully martensitic steels, see also Figure 6.4.

$C_{\mathrm{M}} = 0.6\,\%$. The strain-hardening behavior itself (i.e., the parameters $k_{\mathrm{M}}^{\mathrm{f}}, k_{\mathrm{M}}^{\mathrm{a}}$) is fitted to match results from micro-pillar compression tests.

Experimental results from Jena (1988); Krauss (1999) indicate a linear relation between micro-hardness and carbon contents in the range of $C_{\mathrm{M}} = 0 - 0.7\,\%$. Therefore, it appears to be possible to estimate the martensite strength from experimental hardness results. In the work at hand this is done with hardness values for different carbon contents from the review by Krauss (1999). The estimation is based on the norm for hardness-value conversion (DIN EN ISO NORM DIN EN ISO 18265, 2014). The necessary extrapolation for the estimated yield-strength is supported by nanohardness measurements in Krauss (1999). These hardness´ values stagnate at $\sim$ 10 GPa for carbon-contents above $\sim$ $0.7\,\%$. As a first approximation, hardness or micro-hardness is often assumed to correlate linearly with the yield strength (e.g., Lee et al., 2014; Seok et al., 2014). With the empirical observation of the micro-hardness stagnation for high carbon contents, the martensite yield-stress is estimated to stagnate at $\sigma_{\mathrm{M}}^{\mathrm{f}} \simeq 10\,\mathrm{GPa}/3 = 3.3\,\mathrm{GPa}$.

Note that neither empirical estimations nor (micro-)hardness-based estimations are considered to be exact relations, here. Nevertheless, all empirical estimations and hardness-based estimations are within

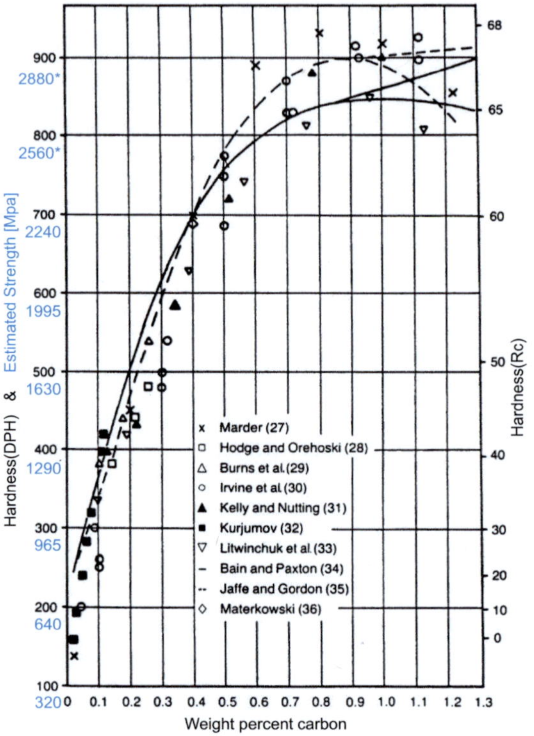

Figure 6.4: Hardness of martensitic microstructures as a function of steel carbon-content (reproduced and modified from Krauss, 1999). Blue values indicate the strength estimation according to the norm for hardness-value conversion (DIN EN ISO NORM DIN EN ISO 18265, 2014); the asterisk indicates linearly extrapolated values. For the martensite of the DP600 steel ($C_{\mathrm{M}} \simeq 0.6 - 0.8\,\%$) one obtains $\sigma^{\mathrm{f}} = 2.3 - 3\,\mathrm{GPa}$ (see Table 6.1 for empirical estimations with $C_{\mathrm{M}} = 0.6 - 0.8\,\%$).

the same reasonable range and, e.g., for the two different DP steels of Chen et al. (2014), the initial yield-strengths of micro-pillar compression tests match the estimated strengths according to Figure 6.4. Additionally, the proposed model and the rule-of-mixture predict an overall correct microscopic and macroscopic behavior of the material for exactly the estimated range. Furthermore, identification of the average martensite constituent behavior based on micro-pillar tests or other micro-scale experiments is difficult.

Gardey et al. (2005) and Haddadi et al. (2006) report absent cross-hardening in DP steels as opposed to interstitial-free steels. The adjustment-rate parameter c was fitted to a subsequent $90°$-tension loading test (uniaxial loading in the rolling direction, unloading, and afterwards uniaxial loading in the transverse direction) for DP600 steel and is employed for all other DP steels, here, too. A redistribution of dislocations after loading path changes is not modeled, here, so the fit for c mainly captures the initial re-yielding characteristics. The employed value of $c = 400$ is close to the value of Tarigopula et al. (2009).

6.8 Algorithmic tangent stiffness

For an application on the Gauss-point level of finite-element simulations, implicit integration-schemes require the algorithmic tangent-stiffness for the global Newton-scheme to be sufficiently correct for rapid convergence. In the following, an approximation is reproduced for the employed mean-field model.

Macroscopic stress is calculated by the ensemble average $\bar{\boldsymbol{\sigma}} = \sum_{\alpha=1}^{N} c_\alpha \boldsymbol{\sigma}_\alpha$. In the special case discussed in Section 2.2, $\Delta\boldsymbol{\sigma}_\alpha = \mathbb{C}_\alpha \left[\Delta\bar{\boldsymbol{\varepsilon}} - \lambda \left\langle \Delta\boldsymbol{\varepsilon}^{\mathrm{p}}_\alpha \right\rangle - (1-\lambda)\, \Delta\boldsymbol{\varepsilon}^{\mathrm{p}}_\alpha\right]$ holds for the microscopic stress in one phase (Equation (2.14)), together with the infinitesimal plastic-strain evolution $\boldsymbol{\varepsilon}^{\mathrm{p}}_\alpha = \boldsymbol{\varepsilon}^{\mathrm{p}}_{\alpha,n} + \Delta\gamma_\alpha \boldsymbol{N}'_\alpha$. With the assumption of a constant plastic interaction, i.e., $\lambda \left\langle \Delta\boldsymbol{\varepsilon}^{\mathrm{p}}_\alpha \right\rangle + (\lambda - 1)\, \Delta\boldsymbol{\varepsilon}^{\mathrm{p}}_\alpha = \mathrm{const.}$, the macroscopic, algorithmic tangent-stiffness is given by

$$\bar{\mathbb{C}}_{\mathrm{alg}} = \sum_{\alpha=1}^{N} c_\alpha \frac{\partial \boldsymbol{\sigma}_\alpha}{\partial \bar{\boldsymbol{\varepsilon}}} = \left\langle \mathbb{C}_{\mathrm{alg}} \right\rangle . \tag{6.40}$$

A closed form solution for $\bar{\mathbb{C}}_{\text{alg}}$ and $\mathbb{C}_{\text{alg}}$ without the assumption $\langle \Delta\varepsilon^{\text{p}}_{\alpha} \rangle = \text{const.}$ is much more elaborated. The main issue in its derivation

$$\bar{\mathbb{C}}_{\text{alg}} = \sum_{\alpha=1}^{N} c_{\alpha} \frac{\partial \Delta \boldsymbol{\sigma}_{\alpha}}{\partial \Delta \bar{\boldsymbol{\varepsilon}}} \tag{6.41}$$

$$= \sum_{\alpha=1}^{N} c_{\alpha} \mathbb{C}_{\alpha} \left(\boldsymbol{I}^{\text{S}} - \lambda \frac{\text{d} \langle \Delta \boldsymbol{\varepsilon}^{\text{p}}_{\alpha} \rangle}{\text{d} \Delta \bar{\boldsymbol{\varepsilon}}} + (\lambda - 1) \frac{\text{d} \Delta \boldsymbol{\varepsilon}^{\text{p}}_{\alpha}}{\text{d} \Delta \bar{\boldsymbol{\varepsilon}}} \right) \tag{6.42}$$

is that for the proposed model (and in the real material) the constitutive models interdepend on the results of the respective other phase. For ferrite one obtains

$$\frac{\text{d}\Delta \boldsymbol{\varepsilon}^{\text{p}}_{\text{F}}}{\text{d}\Delta \bar{\boldsymbol{\varepsilon}}} = \left.\frac{\text{d}\Delta \boldsymbol{\varepsilon}^{\text{p}}_{\text{F}}}{\text{d}\Delta \boldsymbol{\varepsilon}_{\text{F}}}\right|_{\boldsymbol{\varepsilon}_{\text{M}}} \frac{\text{d}\Delta \boldsymbol{\varepsilon}_{\text{F}}}{\text{d}\Delta \bar{\boldsymbol{\varepsilon}}} + \left.\frac{\text{d}\Delta \boldsymbol{\varepsilon}^{\text{p}}_{\text{F}}}{\text{d}\Delta \boldsymbol{\varepsilon}_{\text{M}}}\right|_{\boldsymbol{\varepsilon}_{\text{F}}} \frac{\text{d}\Delta \boldsymbol{\varepsilon}_{\text{M}}}{\text{d}\Delta \bar{\boldsymbol{\varepsilon}}}, \tag{6.43}$$

where $.|_{\phi}$ indicates a constant variable ϕ for the partial derivation. Due to the scaling factor $f_{\text{FM}} = 1 - \dot{\varepsilon}^{\text{p}}_{\text{M}} / \dot{\varepsilon}^{\text{p}}_{\text{F}}$ (Equation (6.33)), the yield direction of ferrite $\boldsymbol{N}'_{\text{F}}$ is actually dependent on the equivalent yield strain of martensite. Thus, the interaction term $\text{d}\Delta\boldsymbol{\varepsilon}^{\text{p}}_{\text{F}} / \text{d}\Delta\boldsymbol{\varepsilon}_{\text{M}}$ can be calculated by utilizing the relation $\Delta\boldsymbol{\varepsilon}^{\text{p}}_{\text{F}} = \Delta\gamma_{\text{F}} \boldsymbol{N}'_{\text{F}} (\dot{\varepsilon}^{\text{p}}_{\text{M}})$,

$$\left.\frac{\text{d}\Delta \boldsymbol{\varepsilon}^{\text{p}}_{\text{F}}}{\text{d}\Delta \boldsymbol{\varepsilon}_{\text{M}}}\right|_{\boldsymbol{\varepsilon}_{\text{F}}} = \left(\left.\frac{\text{d}\Delta \boldsymbol{\varepsilon}^{\text{p}}_{\text{F}}}{\text{d}\Delta \gamma_{\text{F}}}\right|_{\boldsymbol{N}'_{\text{F}}} \otimes \frac{\text{d}\Delta\gamma_{\text{F}}}{\text{d}\Delta \boldsymbol{\varepsilon}_{\text{F}}} \right) \underbrace{\frac{\text{d}\Delta \boldsymbol{\varepsilon}_{\text{F}}}{\text{d}\Delta \boldsymbol{\varepsilon}_{\text{M}}}}_{=0} + \left.\frac{\text{d}\Delta \boldsymbol{\varepsilon}^{\text{p}}_{\text{F}}}{\text{d}\boldsymbol{N}'_{\text{F}}}\right|_{\Delta\gamma_{\text{F}}} \left(\frac{\text{d}\boldsymbol{N}'_{\text{F}}}{\text{d}\Delta\varepsilon^{\text{p}}_{\text{M}}} \otimes \frac{\text{d}\Delta\varepsilon^{\text{p}}_{\text{M}}}{\text{d}\Delta \boldsymbol{\varepsilon}_{\text{M}}} \right). \tag{6.44}$$

The vanishing term is caused by $\boldsymbol{\varepsilon}_{\text{F}} = \text{const.}$ For the calculations in this work, the simple estimation of Equation (6.40) is used for several reasons. Firstly, relatively small time steps must be used for initial changes or load-path changes because both phases strongly interdepend through the evolving long-range stresses. Secondly, the iterative Hashin-Shtrikman mean-field model converges slower with larger time increments for this model with such strong interdependencies.

Finally, all finite-element simulations converged satisfactorily with the simple algorithmic-tangent stiffness. Certainly, the interactive nature of the model results in high computational costs compared to, e.g., simpler Taylor-models with non-interacting materials. This is regarded as the main drawback of the model.

Quantity	Symbol	Unit	DP600	DP980	DF140T	chained 10µmDP	chained 30µmDP	isolated 10µmDP	isolated 30µmDP	CG-DP	FG-DP	UFG-DP
Microstructure (averaged values)												
Ferrite grain size	D_{F}	µm	10	10	8	10	30	10	30	12	2.4	1.2
Mart. grain size	D_{M}	µm	1	5	3.5	3	8	8	20	5	2.4	1.2
Mart.volume fraction	c_{M}	1	0.1	0.6	0.4	0.3	«	«	«	0.3	«	«
Mart. coverage	c_{FM}	1	0.2	1	1	0.9	0.9	0.2	0.3	0.8	0.5	0.5
Strain-partitioning	$(\lambda-1)$	1	$-3\cdot10^{-2}$	$-1\cdot10^{-2}$	$-5\cdot10^{-2}$	$-5\cdot10^{-3}$	$-3\cdot10^{-2}$	$-2\cdot10^{-3}$	$-7\cdot10^{-3}$	$-4\cdot10^{-2}$	$-5\cdot10^{-2}$	$-2\cdot10^{-2}$
Ferrite constituent behavior												
Constant yield-strength	σ_{F}^{0}	MPa	60	270	350	410	280	410	280	300	«	«
Dislocation storage	$k_{\mathrm{F}}^{\mathrm{f}}$	10^9/m	14	1.3	10	2	«	«	«	2	«	«
Dislocation recovery	$k_{\mathrm{F}}^{\mathrm{a}}$	nm	320	200	440	250*	200*	600*	600*	32*	«	«
Initial disl. density	ρ_{F}^{0}	m^{-2}	$1\cdot10^{14}$	$3\cdot10^{11}$	$3\cdot10^{11}$	$2\cdot10^{13}$	$1\cdot10^{13}$	$3\cdot10^{11}$	«	$3\cdot10^{11}$	$1\cdot10^{13}$	$3\cdot10^{13}$
Martensite constituent behavior												
Constant yield-strength	σ_{M}^{0}	MPa	1600	800	1270	1270	787	«	600	1270	1600	1800
Dislocation storage	$k_{\mathrm{M}}^{\mathrm{f}}$	10^9/m	5	20	11	20	«	«	«	11	«	«
Dislocation recovery	$k_{\mathrm{M}}^{\mathrm{a}}$	nm	70	180	150	250	«	«	«	150	«	«
Initial disl. density	ρ_{M}^{0}	m^{-2}	$5\cdot10^{14}$	$1.2\cdot10^{14}$	$8.2\cdot10^{13}$	«	«	«	«	«	«	«
Taylor factor	M	1	3.06	«	«	«	«	«	«	«	«	«
Burgers vector	b	[nm]	0.26	«	«	«	«	«	«	«	«	«
Dislocation interaction	α	1	0.3	«	«	«	«	«	«	«	«	«
Compression modulus	K	GPa	175	«	«	«	«	«	«	«	«	«
Shear modulus	G	GPa	80	«	«	«	«	«	«	«	«	«
Kin. adjustment rate	c	1	400	«	«	«	«	«	«	«	«	«

Table 6.2: Material parameters for a simulated DP steels. See text for references and microstructural based adjustments. Only the values with an asterisk are fitted in a phenomenological manner. The symbol "«" indicates the same value as on the left.

Chapter 7

Results and Discussion

In this section, several applications for the proposed material model are discussed to highlight different effects that occur in DP steels [1]. The model is designed to overcome a severe shortcoming of mean-field models, namely the lack of spatial resolution. However, the underlying physical processes in DP steels are highly localized near the ferrite-martensite interface. The proposed model, thus, aims at reintroducing the governing physical effects on the constituent level in an averaged sense (see Table 6.2 for an overview of all material parameters).

Model results for uniaxial tension and kinematic hardening of DP600 steel are discussed in a general context, at first. As additional applications and validation of the model, two DP steels (DP980, DF140T) from the work of Chen et al. (2014) are considered, due to their detailed characterization. Despite their different microstructures, the two DP980 steels exhibit an identical macroscopic behavior. The model is subsequently applied to four DP steels from the work of Park et al. (2014). Despite having an identical chemical-composition and having equal martensite volume-fractions, the four DP steels highly differ in their microstructure and macroscopic behavior. Finally, the influence of

[1] This section is reproduced from the submitted article: F. Rieger, T. Böhlke, Mechanism based mean-field modeling of the work-hardening behavior of dual-phase steels.

the average ferrite grain-size is discussed with respect to experimental findings of Calcagnotto et al. (2010b).

Note that is not possible to fit a single-phase material-model (a combination of isotropic Voce-hardening and kinematic Armstrong-Frederick-hardening) to the experimental data by (Sebastiano and Brylka, 2014; Sebastiano, 2015). A better fit, instead, is obtained for isotropic hardening as described in Bouaziz (2012) combined with an evolving kinematic saturation-stress. The evolution of the saturation stress (Equation (7.2)) has the an analogous formulation to the one in Haddadi et al. (2006). However, the mentioned model is a purely phenomenological one.

7.1 Uniaxial behavior of dual-phase steel (DP600)

During the loading of a DP steel, different deformation-stages are evident (Korzekwa et al., 1984; Bag and Ray, 1999; Jiang et al., 1995). Deformation stages can be identified from regular Jaoul-Crussard plots ($\ln(\mathrm{d}\sigma/\mathrm{d}\varepsilon)$ vs. $\ln\varepsilon$, see Figure 7.1a) and from modified Jaoul-Crussard plots ($\ln(\mathrm{d}\sigma/\mathrm{d}\varepsilon)$ vs. $\ln\sigma$, see Figure 7.1b). The number of individual stages as well as their characteristics depend on microstructural features (Korzekwa et al., 1984). For example, Mazinani and Poole (2007a) and Han et al. (2011) identify three deformation stages during increasing macroscopic strain: elasticity of both ferrite and martensite, plasticity of ferrite while martensite remains elastic and, finally, plasticity of both constituents, albeit not necessarily at the same yield-rate.

The first transition regime between stage I and stage II hardening behavior has not been thoroughly discussed in the literature about DP steels. Probably, because this transition is less apparent from regular stress-strain plots and only affects the strain-behavior during small initial deformations. The transition from stage II to stage III hardening,

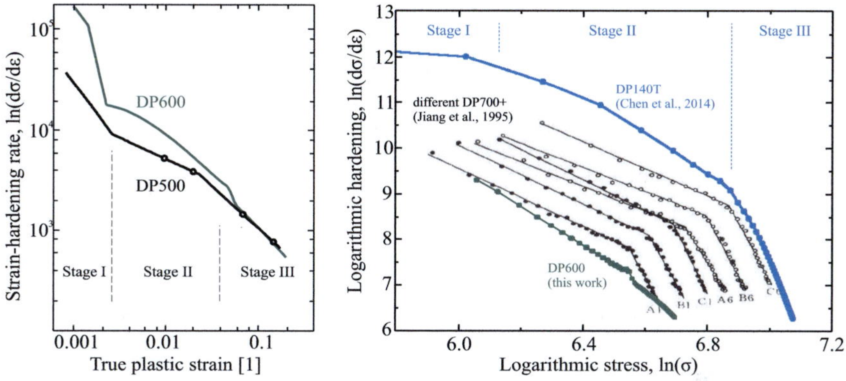

(a) (b)

Figure 7.1: Regular Jaoul-Crussard plot (vs. $\ln \varepsilon$) as modeled and for a DP500 with similar chemical composition (Korzekwa et al., 1984) (a); modified Jaoul-Crussard plot (vs. $\ln \sigma$) for multiple DP700+ steels with higher carbon content and higher $c_{\mathrm{M}} > 0.25$ (Jiang et al., 1995) (b). Colored curves indicate the DP600 (green) and the DF140T (blue) steels from this work. The proposed model is able to predict the different work-hardening stages as they are experimentally observed for the depicted DP steels in the literature.

however, is commonly accepted to be a defining characteristic for DP steels. At deformations of around $\varepsilon \simeq 0.02 - 0.1$, work-hardening exhibits a characteristic kink to a lower-work hardening rate throughout the stage II-stage III transition regime. This kink can be identified in stress-strain plots (see, e.g., Figure 7.3 or Figure 7.4 for demonstrative examples) and in both types of Jaoul-Crussard plots (Figure 7.1a, b). Many authors attribute this kink to beginning plastic-deformation of martensite as it is predicted by the proposed model, see, e.g., the works of Korzekwa et al. (1984); Sarosiek and Owen (1984); Jiang et al. (1995); Bag and Ray (1999).

The three predicted hardening-stages of the DP600 steel are similar to experimental findings of Korzekwa et al. (1984) for a DP500 steel (Figure 7.1a). Regarding Figure 7.1(b), the microstructures of the two color-depicted DP600/DF140T steels represent extremal cases. As a lower bound, DP600 (green) exhibits a small martensite volume-fraction and large ferrite grains. On the other hand, DP980 (blue)exhibits a high martensite volume-fraction and smaller grains which facilitate high work-hardening. The two steels margin the experimental results of Jiang et al. (1995). Their results depict various DP steels with in between microstructures.

7.1.1 Low strain regime – macroscopic elasticity

For low strains, DP steels begin deforming with a macroscopically-elastic deformation-behavior. However, geometry and pre-deformation already lead to inhomogeneous plastic micro-deformation at this stage, e.g., from local stress-concentrations or work-hardened regions (see Chapter 4). The model relies on the single-phase behavior of ferrite with a distinct yield-point. Therefore, it cannot reproduce local plasticity of soft ferrite-regions during the macroscopically-elastic initial deformation. The consecutive high work-hardening due to incompatibility, however, is captured by factoring in geometrically-necessary dislocations.

A small hardening stagnation at the beginning of the transitional regime (Figure 7.2a) has been found in other DP600 steels, too. Colla et al. (2009) attribute this behavior to high phosphor contents of $\sim 0.1\,\%$ that facilitate precipitates. Because the phosphor content is only $0.01\,\%$ in the investigated DP600 steel, this is not a directly applicable explanation, here. However, the EBSD-based dislocation-density estimation for this steel exhibits small point-like areas of high initial densities. The point-like regions could be either very small martensite particles, carbides or other precipitates. Thus, precipitates could, in general, explain a high initial work-hardening in the ferrite constituent.

An experimentally motivated explanation for a hardening stagnation is found in the work of Fonstein et al. (2007). They confirm the common view that suppression of the elongated yield-point phenomenon is facilitated by two effects. Accoring to them, these two effects stem from high internal stresses. These stresses are caused by the austenite-martensite phase-transformation that is restricted by the surrounding ferrite-matrix. Firstly, transformation-induced stresses relax plastically, but still remain present with a significant magnitude, locally adding to the external load. Secondly, plastic relaxation creates mobile dislocations in adjacent ferrite regions. This, locally, allows for an easier accommodation of plastic deformation. Consequently, lower pre-quenching temperatures and lower cooling-rates reduce these thermal stresses, and, thus, can create yield-point elongation. Additionally, the mobile-dislocation density is decreases due to a reduction of the martensite content. Another possibility to reduce the mobile-dislocation density, and,thus, regain elongated yielding, is to increase the carbon content in ferrite, e.g., by tempering. In this case mobile dislocations get pinned by diffusing carbon atoms. Unfortunately, details of the employed thermo-mechanical treatment of the investigated DP600 steel remain unknown. Nevertheless, at least the argument of a low martensite content ($c_{\mathrm{M}} = 0.1$) might be applicable to the present slight yield-point elongation or hardening stagnation which is observed for the DP600 steel.

Another possible explanation can be given in view of the work by Delince et al. (2007). Their model considers long-range stresses from dislocation pile-ups at ferrite-ferrite boundaries. In their work larger ferrite grain-sizes above $5\,\mu\mathrm{m}$ are investigated, similar to the DP600 of this work. For such grain-sizes, their model predicts an overshooting work-hardening contribution from long-range stresses that quickly saturates after a few percent strain. The resulting stress-strain curves look very similar and exhibit a small hardening stagnation directly after initial yielding, too. No such hardening contribution or stagnation is observed for the other investigated DP steels. Therefore, this additional effect is not considered relevant for the work at hand.

7.1.2 Transitional strain-regime – plastic deformation of ferrite

The following deformation step is characteristical for DP steels: exhibiting high work-hardening rates, the low initial yield-stress evolves rapidly towards high yield-strengths (e.g., Sarosiek and Owen (1984)). Two effects are dominant in this regime: Firstly, long-range stresses are created, increasing the bearable external load for ferrite and decreasing the bearable external load for martensite. Secondly, the average dislocation-storage in ferrite is increased by geometrically-necessary incompatibility-dislocations.

Figure 7.2(a) depicts different phase-specific quantities: the single-phase isotropic ferrite yield-stress σ^{f}; the von Mises norm of the long-range (kinematic) stress-tensor $\sqrt{3/2}\,\|\boldsymbol{X}\|$; and, for visualization purposes, the effective one-dimensional yield stress $\sigma^{\mathrm{f}}_{\mathrm{eff}} = \sigma^{\mathrm{f}} \pm \sqrt{3/2}\,\|\boldsymbol{X}\|$ (added for ferrite, subtracted for martensite). Additionally, XRD measurements are shown for the ferrite yield-stress and ferrite long-range stress after unloading the initial load.

Within the DP compound, ferrite shows an initial work-hardening-rate above the single-phase behavior without DP interactions. In the investigated DP600 steel, this effect is not very pronounced because the martensite volume fraction is only $\sim\ 10\,\%$. A more descriptive example for this very high initial-hardening is found in Figure 7.4. In the DP600 steel, however, high initial hardening is present in the fitted ferrite material-behavior. The high initial work-hardening of the ferrite constituent could be explained by means of precipitate-hardening. Precipitate-hardening is not directly considered in the proposed model. Nevertheless, it is indirectly incorporated via the parameter fit for the constituent behavior of ferrite.

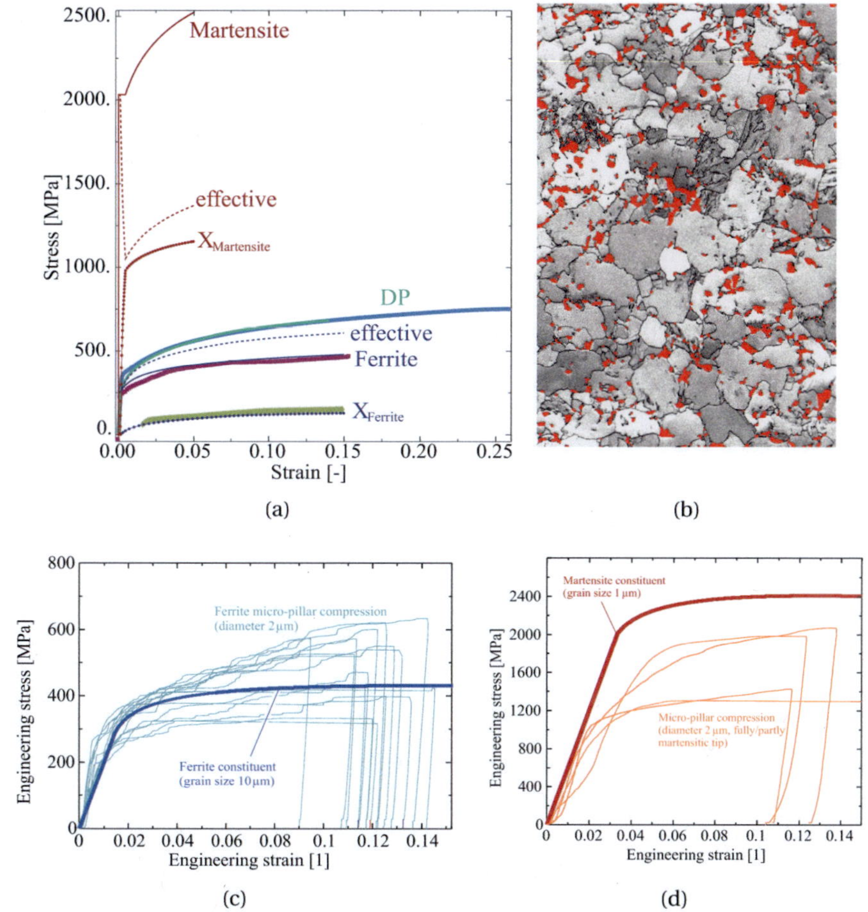

Figure 7.2: Different quantities obtained with the proposed model for the primarily investigated DP600 steel, the predicted macroscopic response is compared to experimental data (a, Schuster (2014); Wenk (2014)); band contrast image of the investigated DP600 steel (b, Wenk, 2014); ferrite (c) and martensite (d) constituent fits in comparison to micro-pillar compression tests (Wenk, 2014). The macroscopic, constituent and long-range stresses are in good agreement with the experimental data, as is the strain-partitioning. Due to the low martensite content, the characteristic kink with two distinctly different work-hardening rates is almost not identifiable in this DP600 steel (a). The red-colored Martensite in (b) is identified by selecting the measurement points with a low band contrast, which do not belong to a grain boundary.

Long-range stresses. The rapid work-hardening is accompanied by a rapid load-transfer, as can be deduced from the decrease of the effective, one-dimensional martensite yield-strength. Load is transferred to martensite by increasing long-range stresses as long as martensite remains elastic and, by this, highly incompatible to ferrite (Ashby, 1970; Brown and Stobbs, 1971a; Mughrabi, 1983). There are only few dislocation-dislocation interactions at small strains when the average dislocation-spacing is large. Therefore, the accumulation of long-range stresses is much faster than the isotropic work-hardening. Following Equation (6.38), the ratio between long-range stresses in ferrite and martensite is fixed by definition. The predicted long-range stresses in ferrite are in good agreement with the XRD measurements during the loading process. This is notable, as the model prediction is solely based on geometrical micromechanical-parameters (Chapter 6). The predicted long-range-stress evolution is also in line with the experimental results of Kim et al. (2003) for a similar DP600 steel ($c_{\mathrm{M}} = 0.12$). Han et al. (2005) also find long-range stresses to increase at least up to a reverse strain of 7 %.

Independent of the microstructural variants of DP steels, Seyedrezai (2014) observes the following for reverse-loading shear experiments: the back-stress itself increases continuously with forward strain up to shear-strains of $\gamma = 0.4$. The *rate* of the back-stress *increase* reduces with forward strain. Experimental data of a DP780 steel from Seyedrezai (2014) suggests that the relative contribution of long-range stresses saturates after $\bar{\varepsilon} = 0.05$. Because overall hardening is still present, this is equivalent to an increase of the long-range stress, similar to the isotropic hardening of the DP steel. In model proposed here, the long-range stress evolution is proportional to the difference in the work-hardening rate between ferrite, and martensite. The strain level of $\bar{\varepsilon} = 0.05$ from the work of Seyedrezai (2014) is slightly lower than in DP600 steel and is associated to the transition from GND dominated work-hardening (stage II) towards dynamic-recovery dominated hardening (stage III). Stage III hardening is not specific for DP steels, but observed for different single-phase metals. The back-stress saturation and the rapid change of the work-hardening rate are observed after equal amounts amount of plastic strain. Therefore, Seyedrezai (2014) concludes, that both effects are caused by similar mechanisms, i.e., plastic deformation and failure of martensite.

The long-range stress magnitude is relatively insensitive to the ferrite work-hardening module. It is predominantly caused by the initial yield-strength of ferrite as the work-hardening rate is always much smaller than the elastic modulus of ferrite. This was also found by Son et al. (2005): load-transfer is higher with higher ferrite yield-strength, leading to an earlier plastic deformation of martensite. Zhonghua and Haicheng (1990) observes similar long-range stress evolutions for different DP steels. The saturation rate and the saturation value at strains of $\bar{\varepsilon} \gtrsim 0.05$ both increase with increasing martensite volume fractions as observed in the work at hand.

As described in Mughrabi (2006), only a small fraction of the GND density actually constitutes long-range stresses for higher deformations. For this DP600 steel, the model presented here yields $\rho_{\mathrm{LRS}} \simeq 1 \cdot 10^{15}\,\mathrm{m}^{-2}$, which is very high. However, the effective long-range stress magnitude is much smaller than the dislocation density ρ_{LRS} from Equation (6.28) would imply. Mughrabi (2006) derives a formula in order to approximate the GND density in the softer phase from the long-range stresses

$$\rho_{\mathrm{GND}}^{\mathrm{LRS}} = \frac{2\left(\sigma_{\mathrm{M}}^{\mathrm{LRS}} - \sigma_{\mathrm{F}}^{\mathrm{LRS}}\right)}{MGbD_{\mathrm{F}}}. \tag{7.1}$$

For the modeled DP600, Equation (7.1) yields values of $\rho_{\mathrm{GND}}^{\mathrm{LRS}} \simeq 2 \cdot 10^{13}\,\mathrm{m}^{-2}$. This is in line with the experimental values of his work, considering the high yield-strength difference in the DP steel: Mughrabi (2006) found $\rho_{\mathrm{GND}}^{\mathrm{LRS}} = 9 \cdot 10^{12}\,\mathrm{m}^{-2}$ for a material with a yield-strength difference of only $80\,\mathrm{MPa}$. The high dislocation-density ρ_{LRS}, however, is not in accordance with Mughrabi (2006). Therefore, it is concluded that ρ_{LRS} cannot be the physically motivated averaged dislocation-density that describes the long-range stress in ferrite. It rather describes the long-range stress that is present in the particles (Equation (6.17)) and at the particles´ interfaces. Recall that the overall dislocation contribution of the incompatibility dislocations scales with c_{M}. The proposed model "distributes" the very-high interfacial stress over the complete ferrite phase (Equation (6.28)). This is certainly only correct in an averaged sense.

The end of the transitional regime of high work-hardening is defined by the beginning of plastic deformation in martensite. There is a characteristic kink in the predicted macroscopic stress-strain behavior. This effect is of minor importance for the DP600 steel due to its low martensite content (see also Figure 7.4). The macroscopic work-hardening rate decreases from this kink on because the work-hardening modulus of martensite is naturally smaller than its elastic modulus. In the transitional regime, martensite exhibits very low strains as its behavior remains elastic within the model. While this is a physical simplification, the local stress-concentrations that cause plastic yielding in martensite are limited and the averaged behavior remains approximately elastic up to several percent of macroscopic strain (Section 4.6). Meanwhile, ferrite bears almost the complete external strain. With the start of plastic deformation in martensite, incompatibilities between ferrite and martensite become less pronounced and the evolution of long-range stresses stagnates. After martensite starts to yield, strain increments become more homogeneously distributed between ferrite and martensite. A changing strain-partitioning during loading is in line with the findings of Shen et al. (1986), Byun and Kim (1993), Thomser et al. (2009) and Chen et al. (2014). They also found ferrite to carry the initial deformation exclusively, until martensite yields and bears a small fraction of the load, too.

7.1.3 High strain regime – plastic deformation of both phases

As soon as both phases deform plastically, incompatibility is certainly reduced. In this regime, the strain-partitioning in both phases becomes more and more equal, i.e., the present strains are then closer to the iso-strain assumption. In consequence, the modeled behavior in this regime is directly and almost exclusively dependent on the behavior of the single phase models because long-range stresses always vanish macroscopically. There is no unified model for single-phase materials produced with different production routes and different chemical compositions, let alone for the constituents in two-phase materials. Especially for the DP600 steel under consideration, this is an issue since micro-pillar compression experiments show a substantial scatter. In essence, for an unknown DP steel, the work-hardening of the high-strain regime must be fit to macroscopic experimental data. Certainly, empirical relations depending on chemical composition and grain size (e.g., Gutiérrez and Altuna, 2008; Resende and Bouvier, 2013) can be used for the constituent parameters σ^0, k^{f} and k^{a} in order for an a priori estimate of a completely unknown DP steel (Section 6.7). This approach, however, is limited and raises issues regarding the quantification of further material parameters. Examples are the Peierls friction-stress σ^0_{PF}, solid solution strengthening σ^0_{SS}, and grain-size effects via σ^0_{GS} and all corresponding interactions.

7.2 Dual-phase steel with martensitic matrix (DP980)

Because the micro-pillar diameter of $\sim 1.5\,\mu\text{m}$ is close to the actual grain-size in the material, the single-constituent fits for ferrite and martensite are in good agreement with the micro-pillar compression tests from Chen et al. (2014) (see Table 6.2 for the material parameters). Strain-partitioning in the DP980 steel is very close to the iso-strain assumption (Chen et al., 2014). This means that the strain-partitioning parameter λ is simple to fit because any value $\lambda \lesssim 0.8$ creates an iso-strain partitioning. The DP980 steel exhibits the two different work-hardening rates that are characteristically for DP steels. An observed initial work-hardening rate up to a macroscopic stress of $\sim 800\,\text{MPa}$ is comparably high with $\sim 100\,\text{GPa}$. Work-hardening decreases rapidly above $\sim 3\,\%$ macroscopic strain and saturates at $\sim 4\,\text{GPa}$. For this DP steel, the ratio between the to work-hardening rates is as big as 200.

The predicted average long-range stresses are about twice as high as in the DP600 steel and they are higher in ferrite than in martensite. This is in line with the finding, that larger martensite volume-fractions are, generally, accompanied by higher long-range stresses. To investigate the influence of the new model contributions, a second simulation for DP980 steel is conducted without any long-range stresses or geometrically-necessary dislocations; the results are practically identical. This insensitivity to long-range stresses is analyzed and explained in the following paragraph.

High work-hardening during the transitional regime is mainly caused by additional forest-dislocation storage due to GNDs (Section 6.3.1). Consequently, a short transitional regime limits the amount of additional dislocation-storage. Three effects determine the length of the transitional regime: the overall strain-partitioning, the difference in initial yield-stresses and the load transfer due to long-range stresses

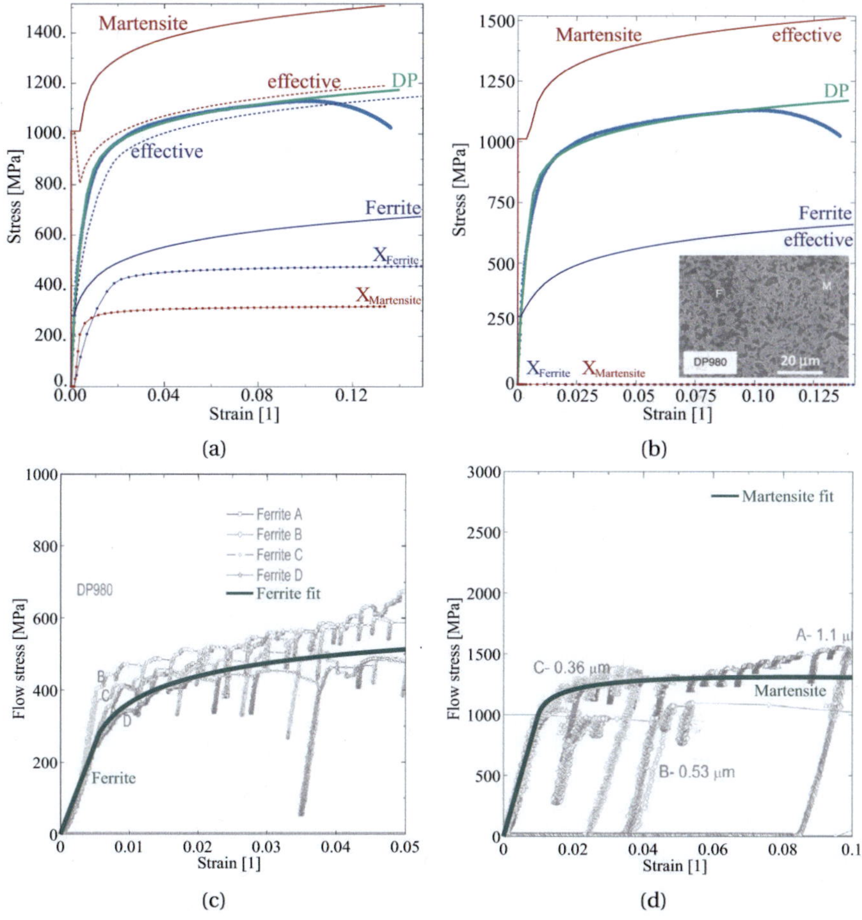

Figure 7.3: Different contributions of the proposed model and macroscopic response compared to experimental data for DP980 steel by Chen et al. (2014) (a); the comparison simulation without incompatibility effects, i.e., no long-range stresses and GNDs (b); ferrite (c) and martensite (d) constituent fits, both compared to micro-pillar compression tests (Chen et al., 2014). Due to exhibiting almost equal strains in both constituents, a simple rule-of-mixture model (b) would suffice to predict the macroscopic behavior. The internal long-range stresses have a negligible influence on the monotonic macroscopic behavior.

(and, additionally, the interaction of all of these effects). One finds the transitional regime in the DP980 steel to be very short, mainly, due to the aforementioned reasons.

For DP steels with a ferrite matrix, the inhomogeneous strain-distribution delays the onset of plasticity in martensite. It is subjected to a smaller fraction of the external strain and ferrite accommodates for this by carrying more strain. Since the end of the transitional regime is determined by the starting plasticity in martensite, this inhomogeneous strain-distribution increases the length of the transitional regime compared to an ideal iso-strain case. For example, Lanzillotto and Pickering (1982) found that martensite remains elastic up to a macroscopic strain of $\bar{\varepsilon} \simeq 0.2$. In the DP980 steel, however, the strain distribution *is* close to the iso-strain case during the complete deformation (Chen et al., 2014). This gives a first reason for the very short transitional regime. Full-field simulations of Chen et al. (2014) support this explanation. They found ferrite to exclusively carry the plastic deformation in the beginning of the deformation. Then, after a very short transitional-regime, the final (iso-)strain-partitioning is reached.

Recall that arbitrary (reasonable) long-range stresses leave the macroscopic model prediction almost unaltered *for monotonic deformations*. On average, they are supposed to cancel each other out by definition. Essentially, the model proposed herein simply predicts the long-range stress saturation X_{sat} in a physically motivated way. Nevertheless, load transfer due to long-range stresses causes martensite to yield even under low external loading. Thus, in DP steels, long-range stresses actually shorten the transitional regime by reducing the difference between the effective yield-stresses of ferrite and martensite. This behavior, in conjunction with the already relatively small yield-strength difference in the DP980 steel, gives another possible reason for the very short observed transitional regime.

High martensite contents. The properties of the DP980 steel are remarkable as it actually exhibits more of a martensitic matrix (60 %) with ferrite inclusions (40 %) than vice versa. By means of the shear-lag analysis, the deformation should be equal in both phases for this case. Thus, there should be no load-transfer via long-range stresses for DP steels with a martensitic matrix (Sarosiek and Owen, 1984; Kim and Lee, 2000). According to Sarosiek and Owen (1984), however, there is a high Bauschinger-effect in DP steels with martensitic matrix. Sarosiek and Owen (1984) state that the high long-range stresses are exclusively constituted by kinematic and isotropic work-hardening mechanisms in martensite that are different from the ones in DP steels with a ferrite matrix. For example, high eigenstresses after the production and the void-like effect of ferrite, together, may govern the behavior of this type of steels. For the long-range stresses in the DP980 steel, up to date, there is no experimental data available in the literature. Thus, a possible contradiction to the shear-lag analysis cannot be examined in detail. Nevertheless, a short comment is, nevertheless, given.

It remains unknown if the effect of highly interconnected martensite for $c_{\mathrm{M}} \simeq 0.5$ is equivalent to a monolithic martensitic matrix as assumed by the shear-lag analysis. For example, Sarwar et al. (2007) found a very similar mechanical behavior for two different DP steels with $c_{\mathrm{M}} \simeq 0.5$: one DP steel exhibits a *continuous* ferrite-matrix with large and isolated martensite (isolated-DP); the other DP steel exhibits *isolated* ferrite-grains with completely interconnected small martensite-grains (connected-DP). The relative martensite-coverage is also quite similar for both DP steels. The connected-DP steel has a microstructure very similar to the investigated DP980 steel, so a closer inspection will give a deeper insight into the topic.

While the martensite in the connected-DP steel is highly interconnected, there is no *monolithic* or coherent martensite matrix. The microstructure rather consists of singular martensite-grains (surrounding ferrite grains) that are in touch via small bottle-necks. Experimental results from Tasan et al. (2014) confirm that regions with high martensite-connectivity are more susceptible to micro-cracking. It seems to be a realistic assumption that martensite does not deform homogeneously, e.g., due to stress-concentrations within these bottlenecks (Tasan et al., 2010) and due to the void-like effect of ferrite (Sarosiek and Owen, 1984). The inhomogeneous deformation of ferrite and martensite, thus, induces long-range stresses that are similar to the modeled ones; at least for equal ferrite/martensite volume-fractions. Indeed, Sarwar et al. (2007) argue that there is (an even more pronounced) stress-transfer from ferrite to martensite in the connected DP steel with $c_{\mathrm{M}} = 0.6$ that is caused by the larger interface area. Note that this, however, contradicts their finding of absent load-transfer for a martensitic matrix. Additionally, Chen et al. (2014), Sarosiek and Owen (1984) and Kim and Lee (2000) all independently conclude that plastic deformation starts in ferrite. This is also found for $c_{\mathrm{M}} > 0.6$ in their work and indicates a behavior akin to the herein proposed model.

It is clear that with a hard martensite matrix instead of a soft ferrite matrix, the model becomes unphysical due to the presence of different mechanisms. Chakraborti and Mitra (2007) observe a kink in yield strength, tensile strength, work-hardening exponent, uniform strain and total strain between $0.6 > c_{\mathrm{M}} > 0.7$ martensite. In view of these experimental findings and the arguments above, the application the herein proposed model to the DP980 steel ($c_{\mathrm{M}} = 0.6$) seems to be a viable extrapolation. Using the proposed model seems not applicable for DP steels with very high martensite volume-fractions above $c_{\mathrm{M}} \gtrsim 0.7$.

7.3 Dual-phase steel with high martensite volume-fractions (DF140T)

As this DP steel has a ferrite matrix, the model is certainly applicable, here. For DF140T, the fitted ferrite-constituent behavior is close to the micro-pillar compression tests (Figure 7.4c). The martensite-constituent behavior, however, deviates (Figure 7.4d). This is, possibly, due to a significant size effect for these smaller grain-sizes: the micro-pillar diameter is $\sim 1\ \mu\mathrm{m}$ but the martensite grain-size is $\sim 4\ \mu\mathrm{m}$ (see Table 6.2 for the material parameters as used in the simulations).

The DF140T steel is a good example for the significance of the contribution from dislocation trapping at ferrite-martensite phase-boundary (β_{FM}) because the predicted ratio $\beta_{\mathrm{MP}}/\beta_{\mathrm{FM}} \simeq 1.5$ is close to unity (Figure 7.4a). The additional effect of the model is illustrated by a comparison simulation without the β-term and without long-range stresses (Figure 7.4b). In order to eliminate the strain-partitioning effect from the comparison, the strain-partitioning parameter λ is adjusted to match the strain distribution of the original simulation. As expected, the comparison simulation shows a similar work-hardening at high strains. The initial work-hardening rate, however, is clearly much smaller if one ignores the incompatibility-contribution; the difference in the initial work-hardening is indicated in red. One can conclude that, although the DF140T steel´s macroscopic behavior is close to that of the DP980 steel, the strengthening mechanisms are different. In the DP980 steel, the high initial work-hardening is mainly caused by a high martensite volume fraction. In the DF140T steel, incompatibility is more pronounced because the martensite-strength is higher due to a higher carbon content. Thus the characteristical high initial work-hardening is caused by additional dislocation-storage due to de-

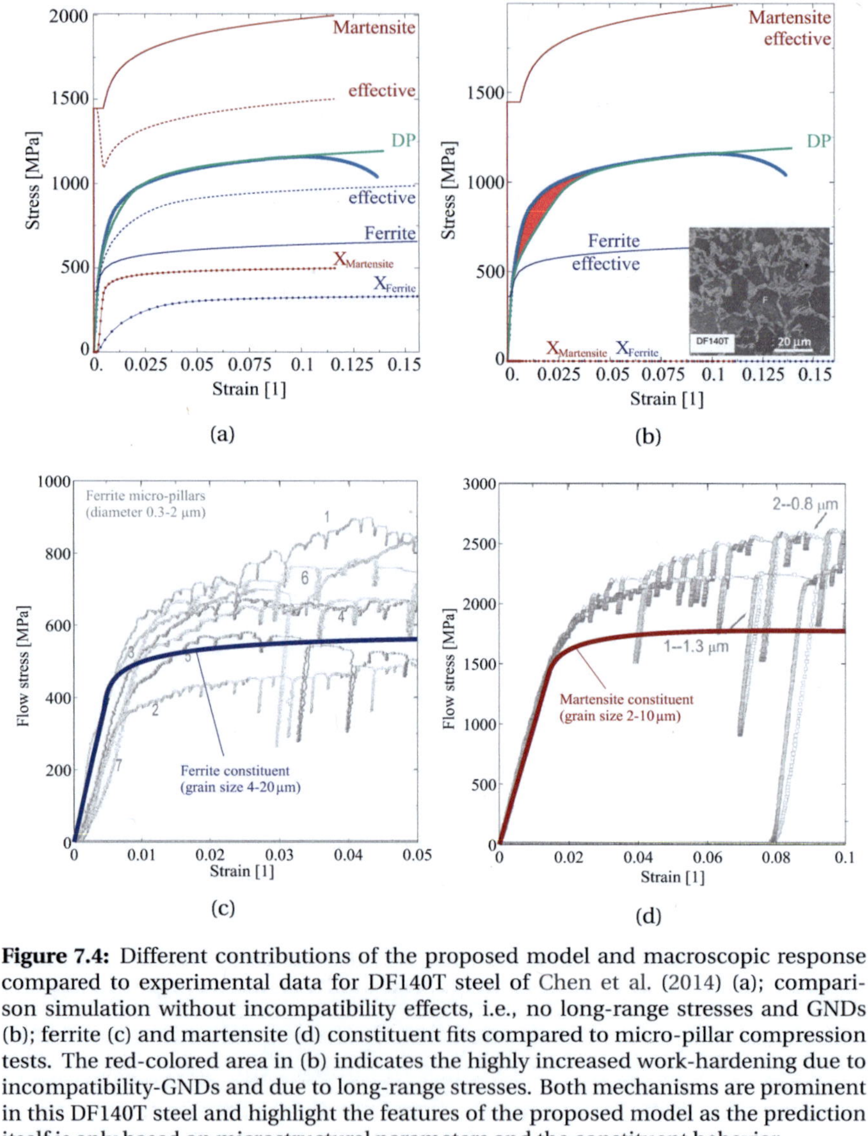

Figure 7.4: Different contributions of the proposed model and macroscopic response compared to experimental data for DF140T steel of Chen et al. (2014) (a); comparison simulation without incompatibility effects, i.e., no long-range stresses and GNDs (b); ferrite (c) and martensite (d) constituent fits compared to micro-pillar compression tests. The red-colored area in (b) indicates the highly increased work-hardening due to incompatibility-GNDs and due to long-range stresses. Both mechanisms are prominent in this DF140T steel and highlight the features of the proposed model as the prediction itself is only based on microstructural parameters and the constituent behavior.

formation incompatibility. This reasoning is in line with the explanation found in the work of Chen et al. (2014).

7.4 Influence of the martensite grain-coverage c_{FM}

Another significant example for the importance of additional geometrically-necessary dislocations from the β_{FM}-term in Equation (6.10) can be found in the work of Park et al. (2014) (Figure 7.5a). There, the amount of ferrite grain-boundaries covered by martensite is identified to be predominant for the work-hardening behavior of the four investigated DP steels. Chemical composition and martensite volume-fractions are identical for all four steels. In Figure 7.5(b), the initial yield-stress (as determined in Park et al., 2014) has already been subtracted from the stresses in the stress-strain curves. This manipulation contracts all curves onto the same yield-point and allows for an isolated comparison of work-hardening mechanisms. Indeed, the initial work-hardening rate is significantly higher for DP steels with a comparably high fraction of ferrite grain-boundaries that are covered by martensite c_{FM}. Compared to the influence of martensite coverage, the effect of the ferrite grain-size on work-hardening is minor for the four DP steels from Park et al. (2014).

Fitting procedure. As an application of the proposed model, all four DP steels from the work of Park et al. (2014) (chained-10DP, chained-30DP, isolated-10DP, isolated-30DP) are reproduced, here. The term "chained" ("isolated") is taken from Park et al. (2014) and indicates a high (low) relative martensite coverage $c_{\mathrm{FM}} \simeq 0.9$ ($c_{\mathrm{FM}} \simeq 0.25$). Unfortunately, there is no experimental data available for the constituents. Thus, the model parameters are fitted with regard to the physical processes that occur in DP steels (see parameters in Table 6.2).

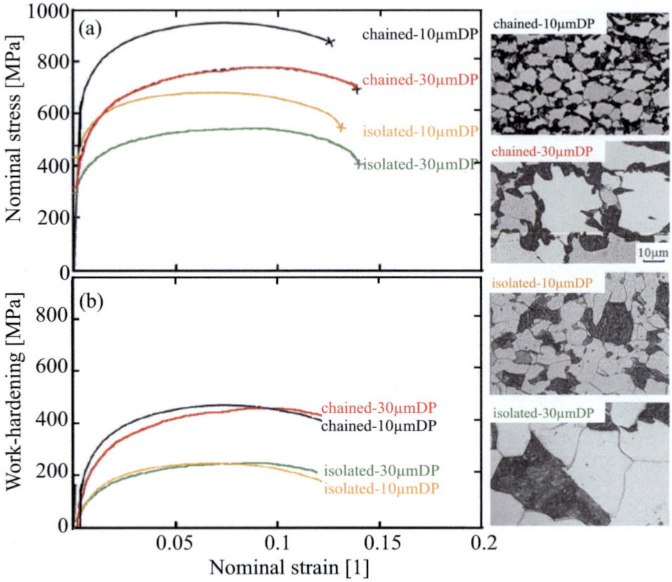

Figure 7.5: Experimental data of four DP steels (a); and comparison of the experimentally determined work-hardening (b). Figures (a) and (b) are reproduced and modified from Park et al., 2014. For all four DP steels in (b), the initial yield stress (as found in the original work of Park et al. (2014)) was subtracted from the original curves to eliminate the grain-size effect and isolate the work-hardening. Figure (b) highlights the predominant influence of relative martensite-coverage compared to grain-size on work-hardening. Additionally, micrographs of the DP steels are reproduced to visualize the different connectivities (isolated-10DP, isolated-30DP, chained-10DP, chained-30DP).

For a ferrite grain-size of 10 μm or 30 μm, both the respective isolated and chained DP steels exhibit similar initial yield-strengths. As plastic deformation starts in ferrite, the initial yield-strength of a DP steel is close to the ferrite yield-strength. The chained-DP steels have a slightly higher initial yield-strength compared to their isolated counterparts with similar grain-sizes. This difference can be readily explained by an increased initial dislocation-density in the chained-DP steels due to the austenite-martensite phase-transformation (Section 4.3). Therefore, the initial dislocation-density is taken to be low in the isolated-DP steels with $\rho_0 = 3 \cdot 10^{11}\,\mathrm{m}^{-2}$, while for the chained-DP steels, $\rho_0 = 1 \cdot 10^{13}\,\mathrm{m}^{-2}$ is assumed. The higher dislocation-density in chained-DP steels is equivalent to an initial yield-strength increase of $\sim 60\,\mathrm{MPa}$ which is in accordance with the experimental data of Park et al. (2014). With regard to these observations, the initial yield-strength σ_{F}^{0} of the respective isolated-DP steels is taken for the ferrite constituent of each of the two $10\,\mu\mathrm{mDP}$, and $30\,\mu\mathrm{mDP}$ steels, respectively. The ferrite grain-size accounts for a difference of $\sim 150\,\mathrm{MPa}$ on the initial yield strength.

Work-hardening without two clearly distinguishable hardening rates indicates less of additional hardening due to incompatibility GNDs and load-transfer. If there was a high amount of additional hardening, the work-hardening rate would suddenly drop as soon as martensite yields, which, in turn, would lead to the characteristical two different hardening rates in DP steels. Regardless of the grain size, chained and isolated DP steels exhibit similar work-hardening between initial and ultimate strength: the isolated-DP steels harden for $\sim 250\,\mathrm{MPa}$ and the chained-DP steels harden for $\sim 450\,\mathrm{MPa}$. The total work-hardening magnitude is probably higher in chained-DP steels because, compared to isolated-DP steels, an increasing amount of ferrite is fully work-hardened (Tasan et al., 2014; Park et al., 2014). Note that Bergström et al. (2010) find the exactly opposite trend in the results from their model. There, similarly to the findings in Chapter 4, a lower amount of hardened ferrite is observed for higher martensite contents and higher martensite coverages. The proposed model in its current state does not account for spatially-heterogeneous dislocation-densities as only an averaged dislocation-density is employed. Consequently, the ferrite work-hardening for high strains has to be adjusted by altering the ferrite dynamic-recovery parameter $k_{\mathrm{F}}^{\mathrm{a}}$, here. The actual dynamic-recovery parameters are identical for both isolated steels and are close to the parameters of the two chained steels. An additional influence on the work-hardening of ferrite emerges from the distinctly different thermo-mechanical production processes for the isolated and chained DP steels. However, it is difficult to quantify this effect without supporting experimental data, and thus to incorporate it into the model that is examined in this section.

Grain-size is taken to be the only influence on the martensite constituent behavior. The martensite grain-size varies between $\sim 3 - 20\,\mu\mathrm{m}$. Findings of Dunstan and Bushby (2014), suggest that the experimental dependence of yield-stress on grain-size is proportional to $1/D$ (or even $\ln(D)/D$), in contrast to the classical Hall-Petch relation $1/\sqrt{D}$. Therefore, σ_{M}^{0} (in MPa) is fitted with $\sigma_{\mathrm{M}}^{0} = 500 + 2300/D$ to account for the high variation of average martensite grain-sizes (D in $\mu\mathrm{m}$). With this ad-hoc assumption the experimental results in (Figure 7.6) can be reproduced.

The strain-partitioning parameter λ is fitted for each DP steel to match the experimental strain-partitioning from Park et al. (2014). Thus, it is not considered to be a "free" parameter as it has been explained above.

Chained-10DP steel. The macroscopic behavior of the chained-10DP steel is similar to the DF140T steel (Figure 7.6). A smaller overall strength can be attributed to the $10\,\%$-reduced martensite volume-fraction and a slightly smaller relative martensite-coverage. There is a deviation in the initial hardening of the predicted stress-strain curve. Since uncertainties are present in the the fitting procedure, several different changes could be made to reduce the deviation. However, this proves to be difficult without additional experimental data.

Chained-30DP steel. Compared to the chained-10DP steel, the chained-30DP steel has a larger average ferrite and martensite grain-size (Figure 7.6). The model should, thus, exhibit two influences.

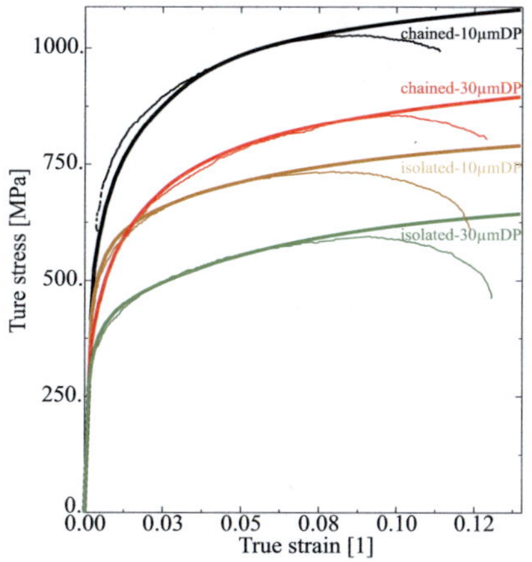

Figure 7.6: Comparison of simulations from this work and the experimental macroscopical behavior from the work of Park et al. (2014): chained-10DP (black); chained-30DP (red); isolated-10DP (yellow); isolated-30DP (green). The utilized parameter fits consider physical mechanisms but could not be verified due to missing experimental data. The macroscopic behavior is in good agreement for all four steels.

Firstly, a reduced overall strength should be modeled because the constant yield-strength parameters are scaled down for both constituents. Secondly, a slightly decreased initial work-hardening rate for the chained-30DP steel should be observable. This effect is caused by the decreasing influence of ferrite and martensite grain-sizes on the additional dislocation storage via $1/D$. In addition, the yield-strength difference is decreased which, in turn, reduces incompatibility between the two phases. As it would be expected, a decreased initial work-hardening rate is clearly noticeable for the chained-30DP steel.

Isolated-10DP steel. Compared to the chained-10DP steel, the macroscopic initial yield-stress of the isolated-10DP steel is only slightly decreased (Figure 7.6). The modeled isolated-10DP steel, again, exhibits good agreement with experimental results. This is remarkable because both simulations use the same initial ferrite work-hardening, and only the saturation strength is altered. The comparison of initial work-hardening for isolated-10DP and chained-10DP steel in Figure 7.5 clearly supports the importance of the additional dislocation-storage with experimental and modeled data.

Isolated-30DP steel. Due to the very large overall average grain-size, the isolated-30DP steel exhibits the lowest yield-stress (Figure 7.6). Work-hardening is almost identical for the isolated-10DP steel because of the rather large grain-size in both isolated DP steels. All β-contributions scale with $1/D$ and their influence on the overall work-hardening is already small for a grain-size of $10\,\mu\mathrm{m}$. The results from the chained-30DP and isolated-30DP steel, again, support the importance of additional dislocation-storage with experimentally determined and modeled data.

For all DP steels from Park et al. (2014), the mechanical behavior is consistently modeled, especially, if the non-existence of experimental constituent data is taken into account. Furthermore, the model is able to reproduce the experimentally measured strain-partitioning for all investigated DP steels in this work (by adjusting λ).

7.5 Influence of the grain size

The effect of an isolated change of the martensite and ferrite grain-size is considered in Calcagnotto et al. (2010b) and also Calcagnotto and Ponge (2009). In comparison to the works of Chen et al. (2014) and Park et al. (2014), one finds a similar chemical composition and martensite volume-fraction ($c_{\mathrm{M}} = 0.3$). Additionally, a high relative martensite-coverage is determined. For a first impression of the isolated grain-size influence, Figure 7.7 depicts a comparison of the tensile data of three DP steels that only differ in their grain sizes (coarse-grained CG-DP, fine-grained FG-DP, ultra-fine-grained UFG-DP). Unfortunately, there

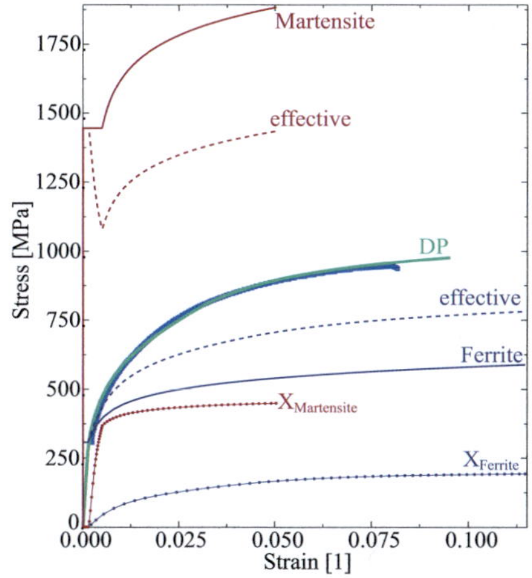

(a)

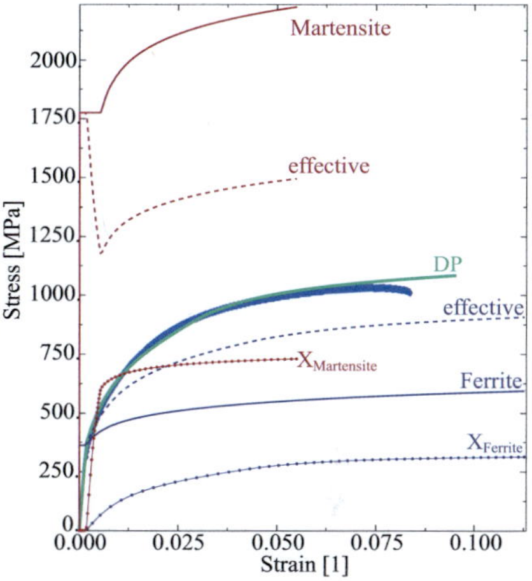

(b)

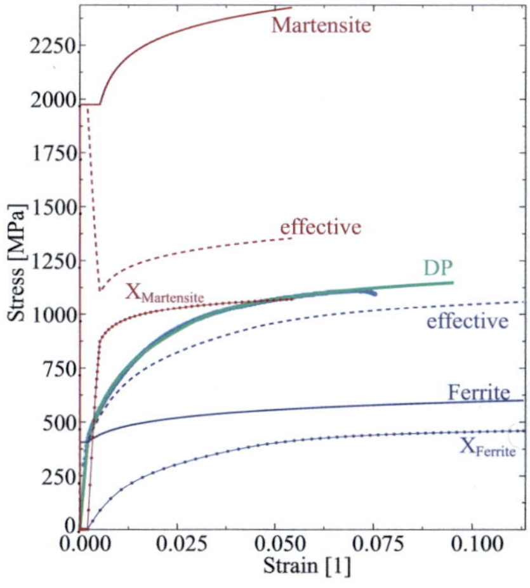

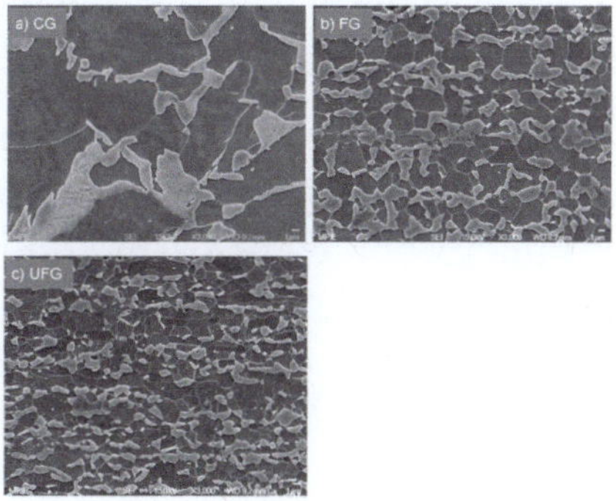

(c) (d)

Figure 7.7: Comparison of experimental data for three DP steels from Calcagnotto et al. (2010b) with the simulations of this work, based on microstructural parameters for martensite: CG-DP (a), FG-DP (b) and UFG-DP (c); reproduced micrographs of the different DP steels (d). The simulated material model is able to reproduce the higher initial work-hardening rate for similar strain-partitioning.

is no experimental data for the constituent phases. Thus, the DF140T steel is taken as basis for determining the CG-DP steel´s parameters as it has a similar chemical composition and mechanical behavior. However, the following adjustments are necessary (Table 6.2). The parameter λ is fitted separately to each DP steel to give a reasonable strain-partitioning in view of the discussion in Section 7.6.

Firstly, the initial ferrite yield-strength σ_{F}^{0} is fitted to the CG-DP steel and kept constant for all simulations. Secondly, the initial dislocation-density ρ_{F}^{0} is set to match the initial yield-strengths from Calcagnotto et al. (2010b). The employed model does not consider ferrite grain-size effects on σ_{F}^{0} but reproduces the experimentally found high dislocation-density throughout the ferrite grains for the FG and UFG steels. Two contributions influence the initial yield-strength: the amount of mobile dislocations and the level of eigenstresses that are created during the phase-transformation. The interaction of these effects makes the quantitative determination of the initial yield strength in DP steels rather complex.

Similar to Section 7.4, grain-size is taken as the only influence on the martensite constituent behavior since the martensite volume-fraction and the carbon content are both equal for all three DP steels. The assumed initial yield-stress of $\sigma_{\mathrm{M}}^{0} \simeq 1200 + 750/D$ (in MPa) accounts for the variation of average martensite grain-sizes D (in $\mu\mathrm{m}$). This phenomenological assumption is primarily taken because the high-strain behavior of all four DP steels can be reproduced with the initial martensite yield-stresses resulting from this equation.

By incorporating the microstructural features, the proposed model is able to reproduce the work-hardening of the three different DP-steels (Figure 7.7a-c). Note that the long-range stress magnitude significantly increases for decreasing grain-sizes. Bergström et al. (2010) find that the ferrite grain-size sensitivity of the initial work-hardening rate is much higher in DP steels than in high-purity iron. This supports the observation that additional, grain-size dependent mechanisms must play a significant role in DP steels. An increasing martensite-strength alone would primarily extend the initial work-hardening rate of CG-DP steel to higher strains but increase the rate itself only slightly. Therefore, the additional hardening mechanisms like martensite particles, and ferrite-martensite dislocation pile-ups, are necessary to model the DP steel behavior. Remarkably, martensite has a very similar effective behavior in all three DP-steels. The higher martensite yield-strength is compensated by higher long-range stresses. Higher long-range stresses are, in turn, directly connected to a higher initial work-hardening rate in the herein presented model.

In accordance with the argument of Section 7.4, the overall influence of grain size seems to be slightly smaller but of similar order compared to the influence of the relative martensite-coverage. For the ferrite grain-sizes ($12, 2.4, 1.2\,\mu\mathrm{m}$) investigated in Calcagnotto et al. (2010a), the initial yield-stress increases by $\sim 80\,\mathrm{MPa}$ while the tensile strength increases by $\sim 160\,\mathrm{MPa}$. Ignoring all other effects, the increase of the relative martensite-coverage in Section 7.4 (Park et al., 2014) increases the tensile strength by $\sim 250\,\mathrm{MPa}$. In the proposed model, the two effects are coupled and both scale proportional to the ratio $c_{\mathrm{FM}}/D_{\mathrm{F}}$.

From the author´s point of view, two questions are crucial for future investigations or extensions of the model. The first one is how to incorporate the spatially-heterogeneous dislocation-density in ferrite. The second question is closely related: how to quantify the effect of the microstructure on the amount of total work-hardening in ferrite (see, e.g, Bergström et al., 2010; Tasan et al., 2014). The clarification and quantification of these effects, would increase the predictive power of the model at hand but also of all mean-field composite models.

7.6 Strain partitioning for different microstructures

Several works indicate the dependence of strain-partitioning on the relative martensite-coverage, e.g., Mazinani and Poole (2007a), Park et al. (2014), and Chen et al. (2014). From experimental data and the shear-lag analysis it is known that for high c_{FM} (a monolithic martensite matrix with soft ferrite inclusions) the iso-strain case is recovered. The other extreme case (rigid inclusions and the soft matrix bears all strain) is a valid assumption for precipitate-like hardening or a highly-plastic matrix with elastic particles. Due to complex microstructural interactions, the transition between the extremal cases is non-trivial in DP-steels. Additionally, for most microstructures, there is be a complex correlation between martensite grain-size, relative martensite-coverage and the shape of the martensitic inclusions. This complicates the evaluation with effective microstructural parameters even further. Strain-partitioning is investigated by means of the microstructural parameters employed in this work. Strain-partitioning data from the end of deformation has been gathered from the literature and is reproduced in Figure 7.8. The data by Mazinani and Poole (2007a) has been obtained within the necking region, which could explain the (only present) outlier (labeled by “0.5?M”).

Clearly, the stress-partitioning changes during the loading process (Byun and Kim, 1993; Thomser et al., 2009; Chen et al., 2014). In the initial small-strain regime, only ferrite accommodates for the macroscopic deformation. This changes after a few percent macroscopic strain: martensite starts to yield and the strain-partitioning changes. For subsequent loading, the strain-partitioning stays almost the same.

For lower martensite volume-fractions, the ratio $r_{\mathrm{M}/DP} = \varepsilon_{\mathrm{M}} / \left(c_{\mathrm{F}}\varepsilon_{\mathrm{F}} + c_{\mathrm{M}}\varepsilon_{\mathrm{M}}\right)$ is small and exhibits a small spread, i.e., martensite bears only small amounts of the average strain (Figure 7.8).

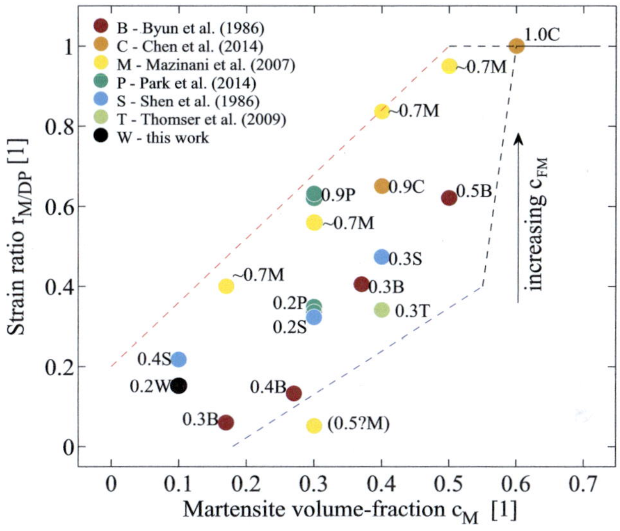

Figure 7.8: Ratio $r_{M/DP}$ of martensite strain ε_M with respect to the total macroscopic strain $\bar{\varepsilon} = c_F\varepsilon_F + c_M\varepsilon_M$, plotted over the martensite volume-fraction. The depicted data is taken from the cited references. Numbers indicate the estimated relative martensite-coverage c_{FM}. As a general trend, one finds that the strain-partitioning becomes more uniform in both phases with an increasing martensite volume-fraction and/or an increasing relative martensite-coverage. For martensite volume-fractions above $c_M \gtrsim 0.6$, the fundamental mechanism changes: DP steel no longer exhibits a ferrite but a monolithic martensite matrix.

With increasing martensite volume-fractions, both the average value and the spread increase (dashed blue and red lines in Figure 7.8). For martensite volume-fractions above $c_M \simeq 0.6$ the fundamental deformation-mechanism changes as DP steel no longer exhibits a ferrite but a martensite matrix. In theory, both phases should undergo the same average deformation for this type of microstructure (dashed black line in Figure 7.8). This explanation is supported by the experimental results of Chakraborti and Mitra (2007), where DP steels exhibit a kink in both yield- and ultimate-stress at precisely $c_M = 0.6$. For such high-martensite DP steels, the proposed model reaches a physical limitation and is no longer applicable.

As a general trend, martensite carries more deformation with a higher relative martensite-coverage. The shear-lag analysis could, again, give an explanation for this behavior: more interfaces enable a higher load transfer. Another consequence of the increased interface surface could be the observed lower bound for the plotted ratio (dashed blue line). With more martensite present in a DP steel, there are probably more interfaces under load. Tasan et al. (2014) also observe that block-structured martensite (low amount of interfaces or low martensite-coverage) leads to higher strain-partitioning. Martensite, consequently, carries much higher stresses at lower strains when compared to ferrite.

The data compiled in Figure 7.8 allows for a first estimation of the strain-partitioning for further simulations. For the presented model, both constituent models and the strain-partitioning parameter λ influence the strain partitioning. The parameter λ is non-linearly related to the strain-partitioning and, therefore, has to be fitted to the experimental strain-partitioning.

7.7 Comparison to kinematic-hardening models from literature

Table 7.1 gives an overview of similar microstructure-based DP models in the literature that are considered relevant for the present work. These works are discussed in the following because all of them highlight a characteristic rapid long-range stress evolution and a high initial dislocation-density production in their modeling.

Reference	c_{M}	D_{F}	**MP/**D_{M}	**FF**	**FM/**c_{FM}
Experiments					
Park et al. (2014)		✓		✓	✓
Sarosiek and Owen (1984)	✓		(✓)		(✓)
Calcagnotto et al. (2010a), Tsipouridis et al. (2011) (initial state)	(✓)	(✓)	(✓)	(✓)	(✓)
Kapp et al. (2011)					(✓)
Mazinani and Poole (2007a), Mazinani and Poole (2007b)	✓		✓		✓
Modeling					
Kim et al. (2012) (3D-RVE)	✓	✓	✓	✓	✓
Ashby (1970) (average dislocation density)	✓		✓	✓	✓
Experiments and modeling					
Jiang et al. (1992), Jiang et al. (1995)	✓	✓	✓	✓	
Zhonghua and Haicheng (1990)	✓		✓		
Sinclair et al. (2006), Delince et al. (2007), Resende and Bouvier (2013)	✓	✓		✓	(✓)
Balliger and Gladman (1981)	✓		✓		
Brown and Stobbs (1971a)/ Lanzillotto and Pickering (1982)	✓		✓		
This work	✓	✓	✓	✓	✓

Table 7.1: Literature overview on the influence of selected microstructural parameters on the work-hardening and long-range stress in DP steels. Here, c_{M} is the martensite volume-fraction, c_{FM} is the fraction of ferrite grain-boundary that is covered by martensite, $D_{\mathrm{F}}/D_{\mathrm{M}}$ are the average ferrite/martensite grain-sizes. The acronyms MP/FF/FM indicate martensite particles, ferrite-ferrite, and ferrite-martensite garin-boundaries, respectively.

In literature, the well-known kinematic-hardening model of Chaboche (1989) has been applied to DP steels with a total kinematic evolution that is constituted by two different "spectral" kinematic-stresses of Frederick-Armstrong type (see Section 6.4.2). The parameters of Tarigopula et al. (2009) are used for the discussion in this paragraph. Both long-range stress contributions in Tarigopula et al. (2009) have a similar saturation value of $\sim 180\,\mathrm{MPa}$ but show evolution rates that differ by an order of magnitude.

In the work at hand, the spectral contribution with the fast evolution rate ($c_1 = 337$) is believed to represent the fast build-up of long-range stresses, i.e., when phases deform highly incompatible and dislocation pile-ups develop rapidly. Recall that incompatibility is at a maximum in this transitional-regime but its evolution almost stops after a few percents of strain; just as this long-range stress contribution does. This correlation has also been identified to be crucial for a mechanism-based explanation of work-hardening by Seyedrezai (2014).The second spectral contribution with a slow evolution rate ($c_2 = 15$) represents two different physical processes for different load cases: During monotonic forward-loading, the second spectral contribution has an effect equivalent to the subsequent increase of X_{sat} (Section 7.1.3). The increase of X_{sat} is caused by the remaining incompatibility of phases after martensite yields; the work-hardening-rate in martensite is often higher than in ferrite. During reverse loading, however, the slow evolution of the second spectral contribution may represent the aforementioned delayed rearrangement of dislocations that leads to a temporary hardening stagnation for reversed loading (as discussed in Section 6.4.2).

Haddadi et al. (2006) utilize different versions of the Hu et al. (1992) model. Essentially, their model version for DP steels predicts an evolving kinematic saturation-stress

$$X_{\mathrm{sat}} = X_0 + S_{\mathrm{D}}, \quad \text{with} \quad \dot{S}_{\mathrm{D}} = C_{\mathrm{SD}} g \left(S_{\mathrm{sat}} - S_{\mathrm{D}} \right) \dot{\gamma}, \quad \text{and} \quad S\left(\varepsilon^{\mathrm{p}} = 0\right) = 0. \tag{7.2}$$

These model relations are similar to the proposed evolution in Equation (6.31). However, they include directional effects with the factor g, describing, e.g., the temporary hardening-stagnation after reverse loading in DP steels. The rate equation for S_{D} has the same structure as the rewritten Armstrong-Frederick law in Equation (6.30), it results in an exponentially saturating kinematic back-stress. Their kinematic-stress tensor saturation ($c = 53$) is about an order of magnitude higher than their saturation rate for X_{sat} ($C_{\text{SD}} = 4$). Note that their model has four parameters (c, X_0, C_{SD}, S_{sat}) to model long-range stresses while the herein proposed material model has only one parameter for the adjustment rate c and none for X_{sat}.

The model of Jiang et al. (1992), that is applied in Jiang et al. (1995), predicts a similar long-range stress behavior. Their model is based on the Ashby (1970) results but neglects the martensite-particle contribution. The derivation is based on an averaged evaluation of dislocation pile-ups at ferrite-ferrite and ferrite-martensite boundaries. As it assumes negligible strain in martensite, the good results only hold for small strains. The model is quite similar to the one proposed in this work, except the authors used more assumptions and less microstructural parameter contributions to obtain an explicit representation of the work hardening.

The long-range stress modeling in the works of Sinclair et al. (2006), Delince et al. (2007) and Resende and Bouvier (2013) are closely connected as they are subsequently derived from each other. The underlying modeling is based on the average interaction of opposing dislocation pile-ups on both sides of single-phase (ferrite-ferrite) grain-boundaries. Interestingly, all of the modeled DP steels have no more than $\sim 50\%$ ferrite-ferrite grain boundaries and it is known that the pile-ups at ferrite-martensite interfaces are more severe (e.g., Dillien et al., 2010; Tsipouridis et al., 2011; Kim et al., 2012). Nevertheless, they predict a long-range stress evolution similar to the work at hand but distinctly require phenomenological parameters to quantify the long-range stress evolution. Their model does neither account for the martensite microstructure, e.g., the observed grain size, or the spatial arrangement of martensite.

Kim et al. (2012) propose different pile-up resistances for ferrite-martensite ($6\,\mathrm{GPa}$) and ferrite-ferrite interfaces ($1.2\,\mathrm{GPa}$). The resulting back-stresses from dislocation pile-ups improve their model predictions significantly. By including such distinct pile-up resistances, they implicitly account for the spatial effect of relative martensite-coverage within their three-dimensional RVE. Essentially, their model exhibits a "perfect-plastic" behavior for the long-range stress at the interfaces. For example, in case of a grain diameter of $5\,\mu\mathrm{m}$, the evolution instantly saturates at a shear strain of $\gamma \simeq 0.05$ for ferrite-ferrite and $\gamma \simeq 0.09$ for ferrite-martensite interfaces. Kim et al. (2012) explicitly state that the highest long-range stresses are located near ferrite-martensite interfaces as the slip activity is more pronounced in these regions. This supports the direct correlation of c_{FM} to the long-range stress evolution in Equation (6.26), and a geometrically-necessary dislocation-density saturation after a few percent of macroscopic strain.

Chapter 8

Conclusion

In the context of grain-size dependent flow-behavior of polycrystals, the following conclusions are drawn in the work at hand:

- For the applied material models, the elastic-plastic transition depends on the strain-partitioning parameter λ but the saturated stress-states do not. The presence of material hardening reduces the influence of λ. In general, stress-fluctuations are, then, decreased within the polycrystal.
- Grain-size dispersion leads to a decrease of the material strength, especially for small mean-diameters around $1\ \mu m$. This effect is caused by the shift of the volume distribution to larger grains with an increasing standard deviation σ_{μ}. Rather than simply the mean grain-size, the whole distribution-shape is important for the macroscopical behavior.
- A representative grain-size, based on the initial yield-stress, allows for a reasonable approximation of the grain-size dispersion effect, at least for isotropic material models. The calculated representative diameter is insensitive within a broad range.

In DP steels, work-hardening is found to mainly depend on the microstructural parameters volume fraction, grain size and relative martensite-coverage. To investigate grain-level plasticity of DP steels, the mechanical behavior of DP steel has been modeled, using an EBSD-based full-field RVE with a virtual quenching step and a subsequent uniaxial-loading step. Ferrite has been modeled by a geometrically-nonlinear crystal-plasticity approach and martensite has been modeled by an isotropic von Mises model. The results were evaluated by means of dislocation-density distributions. The distributions have been analyzed for the initial state (after quenching) and for the evolution with increasing plastic strain:

- EBSD and simulation data indicate that the initial material-state is significantly inhomogeneous and graded from high deformations near ferrite-martensite interfaces to almost unrecognizable deformation inside larger ferrite-grains. In DP steel, there is a grain-size dependent correlation between an increased initial dislocation-density and a higher relative martensite-coverage. At ferrite-ferrite interfaces, almost no increase in dislocation density is observable by EBSD measurements.

- RVE calculations with quenching exhibit a slightly increased yield-point for the macroscopic material-behavior. However, after about $8\,\%$ tensile deformation, the difference between the two simulations vanishes. This highlights the increasing amount of generated plastic-deformation compared to the initial GND-gradient present in dual-phase steels.

- The effect of a higher relative martensite-coverage changes with increasing macroscopic deformation and leads to a decreasing dislocation-density at deformed states above $5\,\%$ strain. A possible explanation is the shielding of ferrite grains from deformation by surrounding martensite.

- Deformation localization appears invariably near martensite islands. However, it is observed that the presence of adjacent martensite-islands does not generally force a high average deformation in ferrite grains at higher strains.

- Initial strain-localization hot-spots exhibit almost unchanged locations but their strain magnitude scales with increasing macroscopic deformation in a self-similar way. The average dislocation-density fluctuation within ferrite grains is about an order of magnitude smaller than the dislocation-density increase from ferrite work-hardening. Therefore, a description of the grain hardening based on the average dislocation-density appears to be a reasonable simplification if one accounts for plastic incompatibility in an averaged sense.

An original material-model for dual-phase steels has been derived, incorporating the interaction of ferrite and martensite via incompatibility-induced long-range stresses in an averaged sense. The proposed model combines works of Ashby (1970) and Brown and Stobbs (1971a) to simulate the ferrite behavior. Based on the composite model of Mughrabi (1983), the constituent interaction is incorporated by a kinematic-hardening contribution. From the simulation results and experimental data is concluded that:

- The ferrite constituent model incorporates an averaged microstructural morphology (i.e., ferrite and martensite volume-fractions, average grain-sizes and relative martensite-coverage). The model-inherent long-range stresses follow a physically motivated evolution law and interact with both ferrite and martensite.

- For the primarily investigated DP steel, the model parameters were identified based on extensive experimental characterizations of the DP600 steel carried out by several co-workers:

uniaxial-tension tests, EBSD measurements, micro-pillar compression tests, and XRD measurements of the ferrite-constituent stresses as well as the long-range stresses in ferrite.

- The two-scale model reproduces the averaged experimental behavior on the grain-level and for the macroscopic behavior of a commercial DP600 steel. Furthermore, the model shows good agreement with experimental data for a range of DP steels from the literature. The influence of the martensite volume-fraction, relative martensite-coverage and grain-size have been examined.
- The constitutive equations comprise a minimal amount of fitting parameters: initial yield-strength, dislocation storage and dislocation recovery for each constituent, the long-range-stress adjustment-rate from the Armstrong-Frederick model, and the strain-partitioning parameter from the mean-field model.
- For high strains, the model response is directly dependent on applicable constituent-models. However, there is no unified model for single-phase materials from different production routes with different chemical compositions, let alone for the averaged constituent behavior in multi-phase materials. The effect of the DP microstructure on the total amount of ferrite work-hardening remains unclear up to this point.
- The proposed model exhibits close similarities to different physically-motivated averaged material-models from the literature. It incorporates the common understanding of physical work-hardening effects in DP steels in a straightforward and physically motivated fashion.

It is noted that the proposed model is not restricted to DP steels, as its foundations have been applied to different metals, too. Possible extensions to the model could separate actively-yielding ferrite areas near the ferrite-martensite interfaces from undeformed regions inside ferrite grains, as previously done in, e.g., Bergström et al. (2010). Additionally, the effect of microstructure on the amount of ferrite work-hardening has not been entirely quantified in the literature, yet.

Bibliography

Asadi, M., De Cooman, B. C., Palkowski, H., Mar. 2012. Influence of martensite volume fraction and cooling rate on the properties of thermomechanically processed dual phase steel. Materials Science and Engineering: A 538, 42–52.

Ashby, M., 1970. The deformation of plastically non-homogeneous materials. Philosophical Magazine 21 (170), 399–424.

Azuma, M., Goutianos, S., Hansen, N., Winther, G., Huang, X., Oct. 2012. Effect of hardness of martensite and ferrite on void formation in dual phase steel. Materials Science and Technology 28 (9), 1092–1100.

Bachmann, F., Hielscher, R., Schaeben, H., Dec. 2011. Grain detection from 2d and 3d EBSD data–specification of the MTEX algorithm. Ultramicroscopy 111 (12), 1720–33.

Bag, A., Ray, K., 1999. Influence of martensite content and morphology on tensile and impact properties of high-martensite dual-phase steels. Metallurgical and Materials Transactions A 30 (May), 1193–1202.

Balliger, N., Gladman, T., 1981. Work hardening of dual-phase steels. Metal Science 15 (March), 95–108.

Barlat, F., Ha, J., Grácio, J. J., Lee, M.-G., Rauch, E. F., Vincze, G., Jul. 2013. Extension of homogeneous anisotropic hardening model to cross-loading with latent effects. International Journal of Plasticity 46, 130–142.

Barlat, F., Lian, K., 1989. Plastic behavior and stretchability of sheet metals. Part I: A yield function for orthotropic sheets under plane stress conditions. International Journal of Plasticity 5, 51–66.

Berbenni, S., Favier, V., Berveiller, M., Jan. 2007. Impact of the grain size distribution on the yield stress of heterogeneous materials. International Journal of Plasticity 23 (1), 114–142.

Bergström, Y., Granbom, Y., Sterkenburg, D., 2010. A Dislocation-Based Theory for the Deformation Hardening Behavior of DP Steels: Impact of Martensite Content and Ferrite Grain Size. Journal of Metallurgy 2010, 1–16.

Bertram, A., 2008. Elasticity and plasticity of large deformations: an introduction, 2nd Edition. Springer, Berlin.

Beynon, N. D., Oliver, S., Jones, T. B., Fourlaris, G., 2005. Tensile and work hardening properties of low carbon dual phase strip steels at high strain rates. Materials Science and Technology 21 (7), 771–778.

Böhlke, T., 2001. Crystallographic Texture Evolution and Elastic Anisotropy: Simulation, Modeling, and Applications. Shaker Verlag.

Böhlke, T., Bondár, G., Estrin, Y., Lebyodkin, M., Feb. 2009. Geometrically non-linear modeling of the Portevin-Le Chatelier effect. Computational Materials Science 44 (4), 1076–1088.

Böhlke, T., Neumann, R., Rieger, F., Jun. 2014. Two-scale modeling of grain size and phase transformation effects. Steel Research International 85 (6), 1018–1034.

Bouaziz, O., 2001. Physical modelling of microstructure and mechanical properties of dual-phase steel. Le Journal de Physiques 11 (PR4), 223–231.

Bouaziz, O., Sep. 2012. Revisited Storage and Dynamic Recovery of Dislocation Density Evolution Law: Toward a Generalized Kocks-Mecking Model of Strain-Hardening. Advanced Engineering Materials 14 (9), 759–761.

Bouaziz, O., Buessler, P., 2002. Mechanical behaviour of multiphase materials: an intermediate mixture law without fitting parameter. Revue de métallurgie, 71–77.

Brown, L. M., Stobbs, W. M., May 1971a. The work-hardening of copper-silica. Philosophical Magazine 23 (185), 1201–1233.

Brown, L. M., Stobbs, W. M., May 1971b. The work-hardening of copper-silica. Philosophical Magazine 23 (185), 1185–1199.

Byun, T. S., Kim, I. S., 1993. Tensile properties and inhomogeneous deformation of ferrite-martensite dual-phase steels. Journal of Materials Science 28 (11), 2923–2932.

Calcagnotto, M., Ponge, D., 2009. Effect of grain refinement on strength and ductility in dual-phase steels. Symposium on Steel (Isss), 1–4.

Calcagnotto, M., Ponge, D., Demir, E., Raabe, D., Apr. 2010a. Orientation gradients and geometrically necessary dislocations in ultrafine grained dual-phase steels studied by 2D and 3D EBSD. Materials Science and Engineering: A 527 (10-11), 2738–2746.

Calcagnotto, M., Ponge, D., Raabe, D., Nov. 2010b. Effect of grain refinement to 1μm on strength and toughness of dual-phase steels. Materials Science and Engineering: A 527 (29-30), 7832–7840.

Carvalho-Resende, T., Balan, T., Bouvier, S., Abed-Meraim, F., Sablin, S.-S., Jan. 2013. Numerical investigation and experimental validation of a plasticity model for sheet steel forming. Modelling and Simulation in Materials Science and Engineering 21 (1), 015008.

Chaboche, J., 1989. Constitutive equations for cyclic plasticity and cyclic viscoplasticity. International Journal of Plasticity 5, 247–302.

Chakraborti, P., Mitra, M., Sep. 2007. Microstructure and tensile properties of high strength duplex ferrite-martensite (DFM) steels. Materials Science and Engineering: A 466 (1-2), 123–133.

Chen, H.-C., Cheng, G.-H., Jun. 1989. Effect of martensite strength on the tensile strength of dual phase steels. Journal of Materials Science 24 (6), 1991–1994.

Chen, P., Ghassemi-Armaki, H., Kumar, S., Bower, A., Bhat, S., Sadagopan, S., Feb. 2014. Microscale-calibrated modeling of the deformation response of dual-phase steels. Acta Materialia 65, 133–149.

Choi, S.-H., Kim, E., Woo, W., Han, S., Kwak, J., Jan. 2013. The effect of crystallographic orientation on the micromechanical deformation and failure behaviors of DP980 steel during uniaxial tension. International Journal of Plasticity 45, 85–102.

Colla, V., De Sanctis, M., Dimatteo, a., Lovicu, G., Solina, a., Valentini, R., Sep. 2009. Strain Hardening Behavior of Dual-Phase Steels. Metallurgical and Materials Transactions A 40 (11), 2557–2567.

Danielsson, M., Parks, D. M., Boyce, M. C., Mar. 2007. Micromechanics, macromechanics and constitutive modeling of the elasto-viscoplastic deformation of rubber-toughened glassy polymers. Journal of the Mechanics and Physics of Solids 55 (3), 533–561.

Davies, R., 1978. Influence of martensite composition and content on the properties of dual phase steels. Metallurgical and Materials Transactions A 9, 671–679.

Dederichs, P. H., Zeller, R., Apr. 1973. Variational treatment of the elastic constants of disordered materials. Zeitschrift fuer Physik 259 (2), 103–116.

Delince, M., Bréchet, Y., Embury, J., Geers, M., Jacques, P., Pardoen, T., Apr. 2007. Structure-property optimization of ultrafine-grained dual-phase steels using a microstructure-based strain hardening model. Acta Materialia 55 (7), 2337–2350.

DeLo, D. P., Semiatin, S. L., 1999. Finite-element modeling of nonisothermal equal-channel angular extrusion. Metallurgical and Materials Transactions A 30 (May), 1391–1402.

Dillien, S., Seefeldt, M., Allain, S., Bouaziz, O., Van Houtte, P., Feb. 2010. EBSD study of the substructure development with cold deformation of dual phase steel. Materials Science and Engineering: A 527 (4-5), 947–953.

Dunstan, D., Bushby, A., 2014. Grain size dependence of the strength of metals: The Hall-Petch effect does not scale as the inverse square root of grain size. International Journal of Plasticity 53, 56–65.

Erdogan, M., Tekeli, S., 2002. The effect of martensite particle size on tensile fracture of surface-carburised AISI 8620 steel with dual phase core microstructure. Materials & design 23 (7), 597–604.

Estrin, Y., 1996. Dislocation-density-related constitutive modeling. In: Krausz, A., Krausz, K. (Eds.), Unified Constitutive Laws of Plastic Deformation. Academic Press, San Diego, pp. 69–106.

Fang, X., Dahl, W., 1995. Strain hardening of steels at large strain deformation. Part I: Relationship between strain hardening and microstructures of bcc steels. Materials Science and Engineering: A 203, 14–25.

Fonstein, N., Kapustin, M., Pottore, N., Gupta, I., Yakubovsky, O., 2007. Factors that determine the level of the yield strength and the return of the yield-point elongation in low-alloy ferrite-martensite steels. The Physics of Metals and Metallography 104 (3), 315–323.

Frederick, C., Armstrong, P., Apr. 2007. A mathematical representation of the multiaxial Bauschinger effect. Materials at High Temperatures 24 (1), 1–26.

Gambin, W., 2002. Plasticity and Textures, 1st Edition. Springer Science+Business Media, Dordrecht.

Gao, S., Chen, M., Chen, S., Kamikawa, N., Shibata, A., Tsuji, N., 2014. Yielding Behavior and Its Effect on Uniform Elongation of Fine Grained IF Steel. Materials Transactions 55 (1), 73–77.

Gardey, B., Bouvier, S., Bacroix, B., 2005. Correlation between the macroscopic behavior and the microstructural evolutions during large plastic deformation of a dual-phase steel. Metallurgical and Materials Transactions A 36 (November), 2937–2945.

Ghassemi-Armaki, H., Maaß, R., Bhat, S., Sriram, S., Greer, J., Kumar, K., Jan. 2014. Deformation response of ferrite and martensite in a dual-phase steel. Acta Materialia 62, 197–211.

Glavas, V., 2014. Personal communication within the DFG Research Training Group 1483, Institute for Engineering Mechanics – Chair for Continuum Mechanics (ITM).

Gross, D., Seelig, T., 2011. Fracture mechanics, 2nd Edition. Springer-Verlag Berlin Heidelberg, Berlin.

Gutiérrez, I., Altuna, M., Oct. 2008. Work-hardening of ferrite and microstructure-based modelling of its mechanical behaviour under tension. Acta Materialia 56 (17), 4682–4690.

Haddadi, H., Bouvier, S., Banu, M., Maier, C., Teodosiu, C., Dec. 2006. Towards an accurate description of the anisotropic behaviour of sheet metals under large plastic deformations: Modelling, numerical analysis and identification. International Journal of Plasticity 22 (12), 2226–2271.

Han, K., Tyne, C. V., Levy, B., 2005. Effect of strain and strain rate on the bauschinger effect response of three different steels. Metallurgical and Materials Transactions A 36 (September), 2379–2384.

Han, Q., Kang, Y., Hodgson, P. D., Stanford, N., Jul. 2013. Quantitative measurement of strain partitioning and slip systems in a dual-phase steel. Scripta Materialia 69 (1), 13–16.

Han, Q.-h., Kang, Y.-l., Zhao, X.-m., Lü, C., Gao, L.-f., May 2011. Microstructure and Properties of Mo Microalloyed Cold Rolled DP1000 Steels. Journal of Iron and Steel Research International 18 (5), 52–58.

He, X., Terao, N., Berghezan, A., 1984. Influence of martensite morphology and its dispersion on mechanical properties and fracture mechanisms of Fe-Mn-C dual phase steels. Metal science.

Hill, R., Hecker, S., Stout, M., 1994. An investigation of plastic flow and differential work hardening in orthotropic brass tubes under fluid pressure and axial load. International Journal of Solids and Structures 31 (21), 2999–3021.

Hu, Z., Rauch, E., Teodosiu, C., 1992. Work-hardening behavior of mild steel under stress reversal at large strains. International Journal of Plasticity 8, 839–856.

Jena, A., 1988. On the effect of the volume fraction on martensite on the tensile strength of dual-phase steel. Materials Science and Engineering, 1–6.

Jiang, Z., Guan, Z., Lian, J., 1995. Effects of microstructural variables on the deformation behaviour of dual-phase steel. Materials Science and Engineering A 190 (1-2), 55–64.

Jiang, Z., Liu, J., Lian, J., Jul. 1992. A new relationship between the flow stress and the microstructural parameters for dual phase steel. Acta Metallurgica et Materialia 40 (7), 1587–1597.

Jöchen, K., 2013. Homogenization of the Linear and Non-linear Mechanical Behavior of Polycrystals. Karlsruher Institut für Technologie (KIT), KIT Scientific Publishing, http://uvka.ubka.uni-karlsruhe.de/shop/download/1000032289.

Jöchen, K., Böhlke, T., Feb. 2012. Prediction of Texture Evolution in Rolled Sheet Metals by Using Homogenization Schemes. Key Engineering Materials 504-506, 649–654.

Kadkhodapour, J., Schmauder, S., Raabe, D., Ziaei-Rad, S., Weber, U., Calcagnotto, M., Jun. 2011. Experimental and numerical study on geometrically necessary dislocations and non-homogeneous mechanical properties of the ferrite phase in dual phase steels. Acta Materialia 59 (11), 4387–4394.

Kang, J., Ososkov, Y., Embury, J., Wilkinson, D., Jun. 2007. Digital image correlation studies for microscopic strain distribution and damage in dual phase steels. Scripta Materialia 56 (11), 999–1002.

Kapp, M., Hebesberger, T., Kolednik, O., Jun. 2011. A micro-level strain analysis of a high-strength dual-phase steel. International Journal of Materials Research (formerly Zeitschrift fuer Metallkunde) 102 (06), 687–691.

Kassner, M., Geantil, P., Levine, L., Nov. 2012. Long range internal stresses in single-phase crystalline materials. International Journal of Plasticity 45, 44–60.

Kiener, D., Motz, C., Schöberl, T., Jenko, M., Dehm, G., Nov. 2006. Determination of Mechanical Properties of Copper at the Micron Scale. Advanced Engineering Materials 8 (11), 1119–1125.

Kim, D., Lee, M., Kim, C., 2003. Measurements of anisotropic yielding, Bauschinger and transient behavior of automotive dual-phase steel sheets. Metals and Materials International 9 (6), 561–570.

Kim, J. H., Kim, D., Barlat, F., Lee, M.-G., Mar. 2012. Crystal plasticity approach for predicting the Bauschinger effect in dual-phase steels. Materials Science and Engineering: A 539, 259–270.

Kim, J.-Y., Jang, D., Greer, J. R., Apr. 2010. Tensile and compressive behavior of tungsten, molybdenum, tantalum and niobium at the nanoscale. Acta Materialia 58 (7), 2355–2363.

Kim, S., Lee, S., 2000. Effects of martensite morphology and volume fraction on quasi-static and dynamic deformation behavior of dual-phase steels. Metallurgical and Materials Transactions A 31 (July), 1753–1760.

Kim, S. a., Johnson, W. L., Apr. 2007. Elastic constants and internal friction of martensitic steel, ferritic-pearlitic steel, and α-iron. Materials Science and Engineering: A 452-453, 633–639.

Kocks, U. F., Mecking, H., 2003. Physics and phenomenology of strain hardening: The FCC case. Progress in Materials Science 48 (3), 171–273.

Koistinen, D. P., Marburger, R. E., 1959. A general equation prescribing the extent of the austenite-martensite transformation in pure iron-carbon alloys and plain carbon steels. Acta Metallurgica 7 (1), 59–60.

Korzekwa, D., Matlock, D., Procedure, E., 1984. Dislocation substructure as a function of strain in a dual-phase steel. Metallurgical and Materials 15 (June), 1221–1228.

Kraft, O., Gruber, P. a., Mönig, R., Weygand, D., Jun. 2010. Plasticity in Confined Dimensions. Annual Review of Materials Research 40 (1), 293–317.

Krauss, G., 1995. Heat treated martensitic steels: microstructural systems for advanced manufacture. ISIJ international 35, 349–359.

Krauss, G., 1999. Martensite in steel: strength and structure. Materials Science and Engineering: A 273-275, 40–57.

Kubin, L., Mortensen, A., 2003. Geometrically necessary dislocations and strain-gradient plasticity: a few critical issues. Scripta Materialia 48 (2), 119 – 125.

Kubin, L. P., Ananthakrishna, G., Fressengeas, C., Franklin, S. V., Marder, M., Mertens, F., 2002. Comment on "Portevin-Le Chatelier effect". Physical Review E - Statistical, Nonlinear, and Soft Matter Physics 65, 3–4.

Kurzydlowski, K. J., 1990. A model for the flow stress dependence on the distribution of grain size in polycrystals. Scripta Metallurgica et Materialia 24 (c), 879–883.

Kuziak, R., Kawalla, R., Waengler, S., Jan. 2008. Advanced high strength steels for automotive industry. Archives of Civil and Mechanical Engineering 8 (2), 103–117.

Lanzillotto, N., Pickering, F. B., 1982. Structure-property relationships in dual-phase steels. Metal Science 16 (August), 371–382.

Larour, P., Bäumer, A., Dahmen, K., Bleck, W., May 2013. Influence of Strain Rate, Temperature, Plastic Strain, and Microstructure on the Strain Rate Sensitivity of Automotive Sheet Steels. Steel Research International 84 (5), 426–442.

Lee, D. J., Yoon, E. Y., Ahn, D.-H., Park, B. H., Park, H. W., Park, L. J., Estrin, Y., Kim, H. S., 2014. Dislocation density-based finite element analysis of large strain deformation behavior of copper under high-pressure torsion. Acta Materialia 76, 281–293.

Liedl, U., Traint, S., Werner, E., 2002. An unexpected feature of the stress-strain diagram of dual-phase steel. Computational materials science 25 (1), 122–128.

Marteau, J., Haddadi, H., Bouvier, S., Jul. 2013. Investigation of Strain Heterogeneities Between Grains in Ferritic and Ferritic-Martensitic Steels. Experimental Mechanics 53 (3), 427–439.

Masumura, R. A., Hazzledine, P. M., Pande, C. S., 1998. Yield stress of fine grained materials. Acta Materialia 46 (13), 4527–4534.

Mazinani, M., Poole, W., Mar. 2007a. Effect of Martensite Plasticity on the Deformation Behavior of a Low-Carbon Dual-Phase Steel. Metallurgical and Materials Transactions A 38 (2), 328–339.

Mazinani, M., Poole, W. J., 2007b. Deformation Behaviour of Martensite in a Low-Carbon Dual-Phase Steel. Advanced Materials Research 15-17, 774–779.

Mecking, H., 2001. Work hardening of single-phase polycrystals. In: Buschow, K. J., Cahn, R. W., Flemings, M. C., Ilschner, B., Kramer, E. J., Mahajan, S. (Eds.), Encyclopedia of Materials - Science and Technology, 2nd Edition. Elsevier, pp. 9785–9795.

Michel, J. C., Suquet, P., 2003. Nonuniform transformation field. International Journal of Solids and Structures2 40, 6937–6955.

Movahed, P., Kolahgar, S., Marashi, S., Pouranvari, M., Parvin, N., Aug. 2009. The effect of intercritical heat treatment temperature on the tensile properties and work hardening behavior of ferrite-martensite dual phase steel sheets. Materials Science and Engineering: A 518 (1-2), 1–6.

Mughrabi, H., 1983. Dislocation wall and cell structures and long-range internal stresses in deformed metal crystals. Acta metallurgica 31 (9), 1367–1379.

Mughrabi, H., 2001. On the role of strain gradients and long-range internal stresses in the composite model of crystal plasticity. Materials Science and Engineering A 317 (1-2), 171–180.

Mughrabi, H., Aug. 2006. Dual role of deformation-induced geometrically necessary dislocations with respect to lattice plane misorientations and/or long-range internal stresses. Acta Materialia 54 (13), 3417–3427.

Mughrabi, H., Ungár, T., Kienle, W., Wilkens, M., Jun. 1986. Long-range internal stresses and asymmetric X-ray line-broadening in tensile-deformed [001]-orientated copper single crystals. Philosophical Magazine A 53 (6), 793–813.

Nadeau, J., Ferrari, M., Nov. 2001. On optimal zeroth-order bounds with application to Hashin-Shtrikman bounds and anisotropy parameters. International Journal of Solids and Structures 38 (44-45), 7945.

Nesterova, E., Bouvier, S., Bacroix, B., 2015. Microstructure evolution and mechanical behavior of a high strength dual-phase steel under monotonic loading. Materials Characterization 100, 152–162.

Nicaise, N., Berbenni, S., Wagner, F., Berveiller, M., Lemoine, X., Feb. 2011. Coupled effects of grain size distributions and crystallographic textures on the plastic behaviour of IF steels. International Journal of Plasticity 27 (2), 232–249.

NORM DIN EN ISO 16842, 2014. Metallic materials - Sheet and strip - Biaxial tensile testing method using a cruciform test piece.

NORM DIN EN ISO 18265, 2014. Metalic Materials - Conversion of hardness values.

NORM DIN EN ISO 6892-1, 2014. Metallic materials - uniaxial tension - Part 1: testing Method of test at room temperature (Appendix B).

Ohser, J., Mücklich, F., 2000. Statistical analysis of microstructures in materials science. Wiley.

Park, K., Nishiyama, M., Nakada, N., Tsuchiyama, T., Takaki, S., May 2014. Effect of the martensite distribution on the strain hardening and ductile fracture behaviors in dual-phase steel. Materials Science and Engineering: A 604, 135–141.

Pipard, J., Nicaise, N., Berbenni, S., 2009. A new mean field micromechanical approach to capture grain size effects. Computational Materials . . . 45 (3), 604–610.

Ponte Castaneda, P., Suquet, P., Castaneda, P. P., 1997. Nonlinear Composites. In: van der Giessen, E., Wu, T. Y. (Eds.), Advances in Applied Mechanics. Vol. 34 of Advances in Applied Mechanics. Elsevier, pp. 171–302.

Queiroz, R., Cunha, F., Gonzalez, B., May 2012. Study of dynamic strain aging in dual phase steel. Materials Science and Engineering: A 543, 84–87.

Raeisinia, B., 2008. Modelling the effect of grain size distribution on the mechanical response of metals. Ph.D. thesis.

Raeisinia, B., Poole, W. J., Jan. 2012. Modelling the elastic-plastic transition of polycrystalline metals with a distribution of grain sizes. Modelling and Simulation in Materials Science and Engineering 20 (1), 015015.

Raeisinia, B., Sinclair, C., Nov. 2009. A representative grain size for the mechanical response of polycrystals. Materials Science and Engineering: A 525 (1-2), 78–82.

Raeisinia, B., Sinclair, C. W., Poole, W. J., Tomé, C. N., Mar. 2008. On the impact of grain size distribution on the plastic behaviour of polycrystalline metals. Modelling and Simulation in Materials Science and Engineering 16 (2), 025001.

Ramazani, A., Mukherjee, K., Schwedt, A., Goravanchi, P., Prahl, U., Bleck, W., Nov. 2013. Quantification of the effect of transformation-induced geometrically necessary dislocations on the flow-curve modelling of dual-phase steels. International Journal of Plasticity 43, 128–152.

Ramtani, S., Bui, H., Dirras, G., Apr. 2009. A revisited generalized self-consistent polycrystal model following an incremental small strain formulation and including grain-size distribution effect. International Journal of Engineering Science 47 (4), 537–553.

Rauch, E., Gracio, J., Barlat, F., May 2007. Work-hardening model for polycrystalline metals under strain reversal at large strains. Acta Materialia 55 (9), 2939–2948.

Rauch, E. F., Gracio, J. J., Barlat, F., Vincze, G., 2011. Modelling the plastic behaviour of metals under complex loading conditions. Modelling and Simulation in Materials Science and Engineering 19 (3), 035009.

Resende, T. C., Bouvier, S., Jan. 2013. Dislocation-based model for the prediction of the behavior of bcc materials - grain size and strain path effects. International Journal of Plasticity 47, 29–48.

Rieger, F., Böhlke, T., Jan. 2015. Microstructure based prediction and homogenization of the strain hardening behavior of dual-phase steel. Archive of Applied Mechanics (special issue), 1–20.

Rodriguez, R.-M., Gutiérrez, I., 2003. Unified Formulation to Predict the Tensile Curves of Steels with Different Microstructures. Materials Science Forum 426-432, 4525–4530.

Sarosiek, A., Owen, W., Sep. 1984. The work hardening of dual-phase steels at small plastic strains. Materials Science and Engineering 66 (1), 13–34.

Sarwar, M., Manzoor, T., Ahmad, E., Hussain, N., Jan. 2007. The role of connectivity of martensite on the tensile properties of a low alloy steel. Materials & Design 28 (6), 1928–1933.

Schmitt, J., Aernoudt, E., Baudelet, B., 1985. Yield loci for polycrystalline metals without texture. Materials Science and Engineering 75, 13–20.

Schmitt, J., Shen, E., Raphanel, J., 1994. A parameter for measuring the magnitude of a change of strain path: Validation and comparison with experiments on low carbon steel. International Journal of Plasticity 10 (5), 535–551.

Schreijäg, S., 2013. Microstructure and Mechanical Behavior of Deep Drawing DC04 Steel at Different Length Scales. Karlsruher Institut für Technologie (KIT), KIT Scientific Publishing, http://uvka.ubka.uni-karlsruhe.de/shop/download/1000032165, Karlsruhe.

Schuster, S., 2014. Personal communication within the DFG Research Training Group 1483, Simone Schuster, Institute for Applied Materials – Materials Science (IAM-WK).

Sebastiano, R., 2015. Experimentelle und numerische Untersuchung von Dualphasenstahl unter biaxialer Belastung, B.Sc. thesis, Karlsruhe Institute of Technology, Institute for Engineering Mechanics – Chair for Continuum Mechanics.

Sebastiano, R., Brylka, B., 2014. Personal communication within the DFG Research Training Group 1483, Institute of Engineering Mechanics – Chair for Continuum Mechanics.

Seok, M.-Y., Kim, Y.-J., Choi, I.-C., Zhao, Y., Jang, J.-i., Aug. 2014. Predicting flow curves of two-phase steels from spherical nanoindentation data of constituent phases: Isostrain method vs. non-isostrain method. International Journal of Plasticity 59, 108–118.

Seyedrezai, H., 2014. Thermo-Mechanical Processing of Dual-Phase Steels and Its Effects on the Work Hardening Behaviour. Thesis, Queens University, Kingston, Ontario, Canada, http://hdl.handle.net/1974/12677.

Shen, H. P., Lei, T. C., Liu, J. Z., 1986. Microscopic deformation behaviour of martensitic-ferritic dual-phase steels. Materials science and technology 2 (January), 28–33.

Simo, J. C., Hughes, T. J. R., 1998. Computational inelasticity. Interdisciplinary applied mathematics ; 7 : Mechanics and materials. Springer, New York.

Sinclair, C., Poole, W., Bréchet, Y., Oct. 2006. A model for the grain size dependent work hardening of copper. Scripta Materialia 55 (8), 739–742.

Sodjit, S., Uthaisangsuk, V., Oct. 2012. Microstructure based prediction of strain hardening behavior of dual phase steels. Materials & Design 41, 370–379.

Son, Y. I., Lee, Y. K., Park, K.-T., Lee, C. S., Shin, D. H., Jun. 2005. Ultrafine grained ferrite-martensite dual phase steels fabricated via equal channel angular pressing: Microstructure and tensile properties. Acta Materialia 53 (11), 3125–3134.

Sung, J. H., Kim, J. H., Wagoner, R., Dec. 2010. A plastic constitutive equation incorporating strain, strain-rate, and temperature. International Journal of Plasticity 26 (12), 1746–1771.

Tang, F., Schoenung, J. M., 2008. Strain softening in nanocrystalline or ultrafine-grained metals: A mechanistic explanation. Materials Science and Engineering A 493, 101–103.

Tarigopula, V., Hopperstad, O., Langseth, M., a.H. Clausen, Jul. 2009. An evaluation of a combined isotropic-kinematic hardening model for representation of complex strain-path changes in dual-phase steel. European Journal of Mechanics - A/Solids 28 (4), 792–805.

Tarigopula, V., Hopperstad, O., Langseth, M., Clausen, A., Sep. 2008. Elastic-plastic behaviour of dual-phase, high-strength steel under strain-path changes. European Journal of Mechanics - A/Solids 27 (5), 764–782.

Tasan, C., Hoefnagels, J., Diehl, M., Yan, D., Roters, F., Raabe, D., Jun. 2014. Strain localization and damage in dual phase steels investigated by coupled in-situ deformation experiments and crystal plasticity simulations. International Journal of Plasticity 63, 198–210.

Tasan, C. C., Hoefnagels, J. J., Geers, M. M., Jun. 2010. Microstructural banding effects clarified through micrographic digital image correlation. Scripta Materialia 62 (11), 835–838.

Thomser, C., Uthaisangsuk, V., Bleck, W., 2009. Influence of Martensite Distribution on the Mechanical Properties of Dual Phase Steels: Experiments and Simulation. Steel Research International 80 (8), 582–587.

Tsipouridis, P., Koll, L., Krempaszky, C., Werner, E., Jun. 2011. On the strength of grain and phase boundaries in ferritic-martensitic dual-phase steels. International Journal of Materials Research (formerly Zeitschrift fuer Metallkunde) 102 (06), 674–686.

Valiev, R., Islamgaliev, R., Alexandrov, I., Mar. 2000. Bulk nanostructured materials from severe plastic deformation. Progress in Materials Science 45 (2), 103–189.

Wagoner, R., Oct. 1981. A new description of strain-rate sensitivity. Scripta Metallurgica 15 (10), 1135–1137.

Wasilkowska, A., Petrov, R., Kestens, L., Werner, E. a., Krempaszky, C., Traint, S., Pichler, a., 2006. Microstructure and Texture Changes in a Low-alloyed TRIP-aided Steel Induced by Small Plastic Deformation. ISIJ International 46 (2), 302–309.

Wenk, M., 2014. Personal communication within the DFG Research Training Group 1483, Institute for Applied Materials – Materials- and Biomechanics (IAM-WBM).

Wenk, M., Schreijäg, S., 2013. Personal communication within the DFG Research Training Group 1483, Institute for Applied Materials – Materials- and Biomechanics (IAM-WBM).

Williams, J., Walters, J., Wang, M., Dec. 2013. Extracting Constitutive Stress-Strain Behavior of Microscopic Phases by Micropillar Compression. JOM 65 (2), 226–233.

Willis, J., 1977. Bounds and self-consistent estimates for the overall properties of anisotropic composites. Journal of the Mechanics and Physics of Solids 25 (3), 185–202.

Wilson, A., Madison, J., Spanos, G., 2001. Determining phase volume fraction in steels by electron backscattered diffraction. Scripta materialia 45, 1335–1340.

Yoon, J. W., Barlat, F., Dick, R. E., Karabin, M. E., 2006. Prediction of six or eight ears in a drawn cup based on a new anisotropic yield function. International Journal of Plasticity 22, 174–193.

Zhonghua, L. I., Haicheng, G. U., 1990. Bauschinger effect and residual phase stresses in two ductile-phase steels: Part II. The effect of microstructure and mechanical properties of the constituent phases on. Metallurgical Transactions A 21 (March), 725–732.

Zhu, B., Asaro, R., Krysl, P., Bailey, R., Oct. 2005. Transition of deformation mechanisms and its connection to grain size distribution in nanocrystalline metals. Acta Materialia 53 (18), 4825–4838.

Zhu, B., Asaro, R., Krysl, P., Zhang, K., Weertman, J., Jul. 2006. Effects of grain size distribution on the mechanical response of nanocrystalline metals: Part II. Acta Materialia 54 (12), 3307–3320.

Schriftenreihe Kontinuumsmechanik im Maschinenbau
Karlsruher Institut für Technologie (KIT)
(ISSN 2192-693X)

Herausgeber: Prof. Dr.-Ing. Thomas Böhlke

Die Bände sind unter www.ksp.kit.edu als PDF frei verfügbar oder als Druckausgabe bestellbar.

Band 1 Felix Fritzen
Microstructural modeling and computational homogenization of the physically linear and nonlinear constitutive behavior of micro-heterogeneous materials. 2011
ISBN 978-3-86644-699-1

Band 2 Rumena Tsotsova
Texturbasierte Modellierung anisotroper Fließpotentiale. 2012
ISBN 978-3-86644-764-6

Band 3 Johannes Wippler
Micromechanical finite element simulations of crack propagation in silicon nitride. 2012
ISBN 978-3-86644-818-6

Band 4 Katja Jöchen
Homogenization of the linear and non-linear mechanical behavior of polycrystals. 2013
ISBN 978-3-86644-971-8

Band 5 Stephan Wulfinghoff
Numerically Efficient Gradient Crystal Plasticity with a Grain Boundary Yield Criterion and Dislocation-based Work-Hardening. 2014
ISBN 978-3-7315-0245-6

Band 6 Viktor Müller
Micromechanical modeling of short-fiber reinforced composites. 2016
ISBN 978-3-7315-0454-2

Schriftenreihe Kontinuumsmechanik im Maschinenbau
Karlsruher Institut für Technologie (KIT)
(ISSN 2192-693X)

Band 7 Florian Rieger
Work-hardening of dual-phase steel. 2016
ISBN 978-3-7315-0513-6